Twisted Pseudodifferential Calculus and Application to the Quantum Evolution of Molecules

MEMOIRS
of the
American Mathematical Society

Number 936

Twisted Pseudodifferential Calculus and Application to the Quantum Evolution of Molecules

André Martinez
Vania Sordoni

July 2009 • Volume 200 • Number 936 (first of 6 numbers) • ISSN 0065-9266

American Mathematical Society
Providence, Rhode Island

2000 *Mathematics Subject Classification.*
Primary 35Q40, 81Q20; Secondary 35S99, 81Q05, 81Q10, 81S30, 81V55.

Library of Congress Cataloging-in-Publication Data

Martinez, André.
Twisted pseudodifferential calculus and application to the quantum evolution of molecules / André Martinez nd Vania Sordoni.
p. cm. — (Memoirs of the American Mathematical Society, ISSN 0065-9266 ; no. 936)
"Volume 200, number 936 (first of 6 numbers)."
Includes bibliographical references and index.
ISBN 978-0-8218-4296-6 (alk. paper)
1. Pseudodifferential operators. 2. Born-Oppenheimer approximation. 3. Wave packets. 4. Evolution equations. 5. Evolution (Biology)—Mathematical models. 6. Quantum theory—Mathematics. I. Sordoni, Vania. II. Title.

QA329.7.M37 2009
515′.7242—dc22 2009008050

Memoirs of the American Mathematical Society

This journal is devoted entirely to research in pure and applied mathematics.

Subscription information. The 2009 subscription begins with volume 197 and consists of six mailings, each containing one or more numbers. Subscription prices for 2009 are US$709 list, US$567 institutional member. A late charge of 10% of the subscription price will be imposed on orders received from nonmembers after January 1 of the subscription year. Subscribers outside the United States and India must pay a postage surcharge of US$65; subscribers in India must pay a postage surcharge of US$95. Expedited delivery to destinations in North America US$57; elsewhere US$160. Each number may be ordered separately; *please specify number* when ordering an individual number. For prices and titles of recently released numbers, see the New Publications sections of the *Notices of the American Mathematical Society.*

Back number information. For back issues see the *AMS Catalog of Publications.*

Subscriptions and orders should be addressed to the American Mathematical Society, P. O. Box 845904, Boston, MA 02284-5904, USA. *All orders must be accompanied by payment.* Other correspondence should be addressed to 201 Charles Street, Providence, RI 02904-2294, USA.

Memoirs of the American Mathematical Society (ISSN 0065-9266) is published bimonthly (each volume consisting usually of more than one number) by the American Mathematical Society at 201 Charles Street, Providence, RI 02904-2294, USA. Periodicals postage paid at Providence, RI. Postmaster: Send address changes to Memoirs, American Mathematical Society, 201 Charles Street, Providence, RI 02904-2294, USA.

This publication is indexed in *Science Citation Index®*, *SciSearch®*, *Research Alert®*, *CompuMath Citation Index®*, *Current Contents®/Physical, Chemical & Earth Sciences.*
Printed in the United States of America.

∞ The paper used in this book is acid-free and falls within the guidelines established to ensure permanence and durability.
Visit the AMS home page at `http://www.ams.org/`

10 9 8 7 6 5 4 3 2 1 14 13 12 11 10 09

Contents

Abstract

We construct an abstract pseudodifferential calculus with operator-valued symbol, suitable for the treatment of Coulomb-type interactions, and we apply it to the study of the quantum evolution of molecules in the Born-Oppenheimer approximation, in the case of the electronic Hamiltonian admitting a local gap in its spectrum. In particular, we show that the molecular evolution can be reduced to the one of a system of smooth semiclassical operators, the symbol of which can be computed explicitely. In addition, we study the propagation of certain wave packets up to long time values of Ehrenfest order.

Received by the editor April 14, 2006.

2000 *Mathematics Subject Classification.* Primary: 35Q40, 81Q20; Secondary 35S99, 81Q05, 81Q10, 81S30, 81V55 .

Key words and phrases. Quantum evolution, Born-Oppenheimer approximation, Pseudodifferential calculus, Operator-valued symbols, Wave-packets.

Investigation supported by University of Bologna. Funds for selected research topics.
Investigation supported by University of Bologna. Funds for selected research topics.

CHAPTER 1

Introduction

In quantum physics, the evolution of a molecule is described by the initial-value Schrödinger system,

$$(1.1) \qquad \begin{cases} i\partial_t\varphi = H\varphi; \\ \varphi\,|_{t=0} = \varphi_0, \end{cases}$$

where φ_0 is the initial state of the molecule and H stands for the molecular Hamiltonian involving all the interactions between the particles constituting the molecule (electron and nuclei). In case the molecule is imbedded in an electromagnetic field, the corresponding potentials enter the expression of H, too. Typically, the interaction between two particles of positions z and z', respectively, is of Coulomb type, that is, of the form $\alpha|z-z'|^{-1}$ with $\alpha \in I\!R$ constant.

In the case of a free molecule, a first approach for studying the system (1.1) consists in considering bounded initial states only, that is, initial states that are eigenfunctions of the Hamiltonian after removal of the center of mass motion. More precisely, one can split the Hamiltonian into,

$$H = H_{\rm CM} + H_{\rm Rel},$$

where the two operators H_{CM} (corresponding to the kinetic energy of the center of mass) and H_{Rel} (corresponding to the relative motion of electrons and nuclei) commute. As a consequence, the quantum evolution factorizes into,

$$e^{-itH} = e^{-itH_{\rm CM}}e^{-itH_{\rm Rel}},$$

where the (free) evolution $e^{-itH_{\rm CM}}$ of the center of mass can be explicitly computed (mainly because $H_{\rm CM}$ has constant coefficients), while the relative motion $e^{-itH_{\rm Rel}}$ still contains all the interactions (and thus, all the difficulties of the problem). Then, taking φ_0 of the form,

$$(1.2) \qquad \varphi_0 = \alpha_0 \otimes \psi_j$$

where α_0 depends on the position of the center of mass only, and ψ_j is an eigenfunction of $H_{\rm Rel}$ with eigenvalue E_j, the solution of (1.1) is clearly given by,

$$\varphi(t) = e^{-itE_j}(e^{-itH_{\rm CM}}\alpha_0) \otimes \psi_j.$$

Therefore, in this case, the only real problem is to know sufficiently well the eigenelements of H_{Rel}, in order to be able to produce initial states of the form (1.2).

In 1927, M. Born and R. Oppenheimer [**BoOp**] proposed a formal method for constructing such an approximation of eigenvalues and eigenfunctions of H_{Rel}. This method was based on the fact that, since the nuclei are much heavier than the electrons, their motion is slower and allows the electrons to adapt almost instantaneously to it. As a consequence, the motion of the electrons is not really perceived by the nuclei, except as a surrounding electric field created by their total potential energy (that becomes a function of the positions of the nuclei). In that way, the evolution of the molecule reduces to that of the nuclei imbedded in an effective electric potential created by the electrons. Such a reduction (that is equivalent to a decomposition of the problem into two different position-scales) permits, in a second step, to use semiclassical tools in order to find the eigenelements of the final effective Hamiltonian.

At this point, it is important to observe that this method was formal only, in the sense that it allowed to produce formal series of functions that were (formally) solutions of the eigenvalue problem for H_{Rel}, but without any estimates on the remainder terms, and no information about the possible closeness of these functions to true eigenfunctions, nor to the possible exhaustivity of such approximated eigenvalues.

Many years later, a first attempt to justify rigorously (from the mathematical point of view) the Born-Oppenheimer approximation (in short: BOA) was made by J.-M. Combes, P. Duclos and R. Seiler [**CDS**] for the diatomic molecules, with an accuracy of order h^2, where $h := \sqrt{m/M}$ is the square-root of the ratio of the electron masses to nuclear masses. After that, full asymptotics in h were obtained by G. Hagedorn [**Ha2, Ha3**], both in the case of diatomic molecules with Coulomb interactions, and in the case of smooth interactions. In these two cases, these results gave a positive answer to the first question concerning the justification of the BOA, namely, the existence of satisfactory estimates on the remainder terms of the series. Later, by using completely different methods (mostly inspired by the microlocal treatment of semiclassical spectral problems, developed by B. Helffer and J. Sjöstrand in [**HeSj11**]), and in the case of smooth interactions, the first author [**Ma1**] extended this positive answer to the two remaining questions, that is, the exhaustivity and the closeness of the formal eigenfunctions to the true ones. Although such a method (based on microlocal analysis) seemed to require a lot of smoothness, it appeared that it could be adapted to the case of Coulomb interactions, too, giving rise to a first complete rigorous justification of the BOA in a work by M. Klein, A. Martinez, R. Seiler and X.P. Wang [**KMSW**]. The main trick, that made possible such an adaptation, consists in a change of variables in the positions of the electrons, that depends in a convenient way of the position (say, x) of the nuclei. This permits to make the singularities of the interactions electron-nucleus independent of x, and thus, in some sense, to regularize these interactions with respect to x. Afterwards, the standard microlocal tools (in particular, the pseudodifferential calculus with operator-valued symbols, introduced in [**Ba**]) can be applied and give the conclusion.

Of course, all these justifications concerned the eigenvalue problem for H_{Rel}, not the general problem of evolution described in (1.1). In the general case, one could think about expanding any arbitrary initial state according to the eigenfunctions of H_{Rel}, and then apply the previous constructions to each term. However, this would lead to remainder terms quite difficult to estimate with respect to the small parameter h, mainly because one would have to mix two types of approximations that have nothing to do with each other: The semiclassical one, and the eigenfunctions expansion one. In other words, this would correspond to handle both functional and microlocal analysis, trying to optimize both of them at the same time. It is folks that such a method is somehow contradictory, and does not produce good enough estimates. For this reason, several authors have looked for an alternative way of studying (1.1), by trying to adapt Born-Oppenheimer's ideas directly to the problem of evolution.

The first results in this direction are due to G. Hagedorn [**Ha4, Ha5, Ha6**], and provide complete asymptotic expansions of the solution of (1.1), in the case of smooth interactions and when the initial state is a convenient perturbation of a single electronic-level state. More precisely, splitting the Hamiltonian into,

$$H = K_{\rm n}(hD_x) + H_{\rm el}(x),$$

where $K_{\rm n}(hD_x)$ stands for the quantum kinetic energy of the nuclei, and $H_{\rm el}(x)$ is the so-called electronic Hamiltonian (that may be viewed as acting on the position variables y of the electrons, and depending on the position x of the nuclei), one assumes that $H_{\rm el}(x)$ admits an isolated eigenvalue $\lambda(x)$ (say, for x in some open set of $I\!R^3$) with corresponding eigenfunction $\psi(x,y)$, and one takes φ_0 of the form,

$$\varphi_0(x,y) = f(x)\psi(x,y) + \sum_{k\geq 1} h^k \varphi_{0,k}(x,y) = f(x)\psi(x,y) + \mathcal{O}(h),$$

where $f(x)$ is a coherent state in the x-variables. Then, it is shown that, if the $\varphi_{0,k}$'s are conveniently chosen, the solution of (1.1) (with a rescaled time $t \mapsto t/h$) admits an asymptotic expansion of the type,

$$\varphi_t(x,y) \sim f_t(x)\psi(x,y) + \sum_{k\geq 1} h^k \varphi_{t,k}(x,y),$$

where all the terms can be explicitly computed by means of the classical flow of the effective Hamiltonian $H_{\rm eff}(x,\xi) := K_{\rm n}(\xi) + \lambda(x)$.

Such a result is very encouraging, since it provides a case in which the relevant information on the initial state is not anymore connected with the point spectrum of $H_{\rm rel}$, but rather with the localization in energy of the electrons and the localization in phase space of the nuclei. This certainly fits much better with the semiclassical intuition of this problem, in accordance with the fact that the classical flow of $H_{\rm eff}(x,\xi)$ is involved.

Nevertheless, from a conceptual point of view, something is missing in the previous result. Namely, one would like to have an even closer relation between the complete quantum evolution $e^{-itH/h}$ and some *reduced quantum evolution* of the type $e^{-it\tilde{H}_{\rm eff}(x,hD_x)/h}$, for some $\tilde{H}_{\rm eff}$ close to $H_{\rm eff}$. In that way, one would be able to use all the well developed semiclassical (microlocal) machinery on the operator $\tilde{H}_{\rm eff}(x,hD_x)$, in order to deduce many results on its quantum evolution group $e^{-it\tilde{H}_{\rm eff}(x,hD_x)/h}$ (e.g., a representation of it as a Fourier integral operator). In the previous result, the presence of a coherent state in the expression of φ_0 has allowed the author to, somehow, by-pass this step, and to relate directly the complete quantum evolution to its semiclassical approximation (that is, to objects involving the underlying classical evolution). However, a preliminary link between $e^{-itH/h}$ and some $e^{-it\tilde{H}_{\rm eff}(x,hD_x)/h}$ would have the advantage of allowing more general initial states, and, by the use of more sophisticated results of semiclassical analysis, should permit to have a better understanding of the phenomena related to this approximation. Moreover, as we will see, this preliminary link is usually valid for very large time intervals of the form $[-h^{-N}, h^{-N}]$ with $N \geq 1$ arbitrary, while it is well known that the second step (that is, the semiclassical approximation of $e^{-it\tilde{H}_{\text{eff}}(x,hD_x)/h}$) has, in best cases, the Ehrenfest-time limitation $|t| = \mathcal{O}(\ln \frac{1}{h})$ (see (2.5) and Theorem 11.3 below).

The first results concerning a reduced quantum evolution have been obtained recently (and independently) by H. Spohn and S. Teufel in [**SpTe**], and by the present authors in [**MaSo**]. In both cases, it is assumed that, at time $t = 0$, the energy of the electrons is localized in some isolated part of the electronic Hamiltonian $H_{\text{el}}(x)$. In [**SpTe**], the authors find an approximation of $e^{-itH/h}$ in terms of $e^{-itH_{\text{eff}}(x,hD_x)/h}$, and prove an error estimate in $\mathcal{O}(h)$ (actually, it seems that such a result was already present in a much older, but unpublished, work by A. Raphaelian [**Ra**]). In [**MaSo**] (following a procedure of [**NeSo, So**], and later reproduced with further applications in [**PST, Te**]), a whole perturbation $\tilde{H}_{\text{eff}} \sim H_{\text{eff}} + \sum_{k\geq 1} h^k H_k$ of H_{eff} is constructed, allowing an error estimate in $\mathcal{O}(h^\infty)$ for the quantum evolution.

However, these two papers have the defect of assuming all the interactions smooth, and thus of excluding the physically interesting case of Coulomb interactions. Here, our goal is precisely to allow this case. More precisely, we plan to mix the arguments of [**MaSo**] and those of [**KMSW**] in order to include Coulom-type (or, more generally, Laplace-compact) singularities of the potentials.

In [**KMSW**], the key-point consists in a refinement of the Hunziker distorsion method, that leads to a family of x-dependent unitary operators (where, for each operator, the nuclei-position variable x has to stay in some small open set) such that, once conjugated with these operators, the electronic Hamiltonian becomes smooth with respect to x. Then, by using local pseudodifferential calculus with operator-valued symbols, and various tricky patching techniques, a constructive Feshbach method (through a Grushin problem) is performed and leads to the required result.

When reading [**KMSW**], however, one has the impression that all the technical difficulties and tricky arguments actually hide a somewhat simpler concept, that should be related to some global pseudodifferential calculus adapted to the singularities of the interactions. In other words, it seems that interactions such as Coulomb electron-nucleus ones are, indeed, smooth with respect to x for some 'exotic' differential structure on the x-space, and that such a differential structure could be used to construct a complete pseudodifferential calculus (with operator-valued symbols). Such considerations (that are absent in [**KMSW**]) have naturally led us to the notion of *twisted pseudodifferential operator* that we describe in Capters 4 and 5. This new tool permits in particular to handle a certain type of partial differential operators with singular operator-valued coefficients, mainly as if their coefficients were smooth. To our opinion, the advantages are at least two. First of all, it simplifies considerably (making them clearer and closer to the smooth case) the arguments leading to the reduction of the quantum evolution of a molecule. Secondly, thanks to its abstract setting, we believe that it can be applied in other situations where singularities appear.

Roughly speaking, we say that an operator P on $L^2(\mathbb{R}^n_x; \mathcal{H})$ ($\mathcal{H}$ = abstract Hilbert space) is a twisted h-admissible pseudodifferential operator, if each operator $U_j P U_j^{-1}$ (where, for any j, $U_j = U_j(x)$ is a given unitary operator defined for x

in some open set $\Omega_j \subset I\!R^n$) is h-admissible (e.g., in the sense of [**Ba, GMS**]). Then, under few general conditions on the finite family $(U_j, \Omega_j)_j$, we show that these operators enjoy all the nice properties of composition, inversion, functional calculus and symbolic calculus, similar to those present in the smooth case. Thanks to this, the general strategy of [**MaSo**] can essentially be reproduced, and leads to the required reduction of the quantum evolution. More precisely, we prove that, if the initial state φ_0 is conveniently localized in space, in energy, and on a L-levels isolated part of the electronic spectrum ($L \geq 1$), then, during a certain interval of time (that can be estimated), its quantum evolution can be described by that of a selfadjoint $L \times L$ matrix $A = A(x, hD_x)$ of smooth semiclassical pseudodifferential operators in the nuclei-variables, in the sense that one has,

$$e^{-itH/h}\varphi_0 = \mathcal{W}^* e^{-itA/h}\mathcal{W}\varphi_0 + \mathcal{O}(\langle t\rangle h^\infty),$$

where $\mathcal{W}$ is a bounded operator onto $L^2(I\!R^n)^{\oplus L}$, such that $\mathcal{W}\mathcal{W}^* = 1$ and $\mathcal{W}^*\mathcal{W}$ is an orthogonal projection (that projects onto a so-called almost-invariant subspace). We refer to Theorem 2.1 for a precise statement, and to Theorem 7.1 for an even better result in the case where the spectral gap of the electronic Hamiltonian is global. In the particular case $L = 1$, this also permits to give a geometrical description (involving the underlying classical Hamilton flow of A) of the time interval in which such a reduction is possible. Then, to make the paper more complete, we consider the case of coherent initial states (in the same spirit as in [**Ha5, Ha6**]) and, applying a semiclassical result of M. Combescure and D. Robert [**CoRo**], we justify the expansions given in [**Ha6**] up to times of order $\ln \frac{1}{h}$ (at least when the geometry makes it possible).

Outline of the paper:

In Chapter 2, we introduce our notations and assumptions, and we state our main results concerning the reduction of the quantum evolution in the case where the electronic Hamiltonian admits a local gap in its spectrum. In Chapter 3, we modify the electronic operator away from the relevant region in x, in order to deal with a globally nicer operator, admitting a global gap in its spectrum. Chapters 4 and 5 are devoted to the settlement of an abstract singular pseudodifferential calculus (bounded in Chapter 4, and partial differential in Chapter 5). In Chapter 6, following [**MaSo**], we construct a quasi-invariant subspace that permits, in Chapter 7, to have a global reduction of the evolution associated with the modified operator constructed in Chapter 3. In Chapters 8 and 9, we complete the proofs of our main results, and, in Chapter 10, we give a simple way of computing the effective Hamiltonian. Then, in Chapter 11, we apply these results to study the evolution of wave packets. Chapter 12 treats, more specifically, the case of polyatomic molecules, by showing how it can be inserted in our general framework. The remaining three chapters are just appendices: Chapter A reviews standard results on pseudodifferential calculus; Chapter B gives an estimate on the propagation-speed of the support (up to $\mathcal{O}(h^\infty)$) of the solutions of (1.1); Chapter C contains two technical results used in the paper.

CHAPTER 2

Assumptions and Main Results

The purpose of this paper is to investigate the asymptotic behavior as $h \to 0_+$ of the solutions of the time-dependent Schrödinger equation,

$$(2.1)\qquad ih\frac{\partial\varphi}{\partial t} = P(h)\varphi$$

with

$$(2.2)\qquad P(h) = \boldsymbol{\omega} + Q(x) + W(x),$$

where $Q(x)$ $(x \in I\!R^n)$ is a family of selfadjoint operators on some fix Hilbert space $\mathcal{H}$ with same dense domain $\mathcal{D}_Q$, $\boldsymbol{\omega} = \sum_{|\alpha|\leq m} c_\alpha(x;h)(hD_x)^\alpha$ is a symmetric semiclassical differential operator of order 0 and degree m, with scalar coefficients depending smoothly on x, and $W(x)$ is a non negative function defined almost everywhere on $I\!R^n$.

Typically, in the case of a molecular system, x stands for the position of the nuclei, $Q(x)$ represents the electronic Hamiltonian that includes the electron-electron and nuclei-electron interactions (all of them of Coulomb-type), $\boldsymbol{\omega}$ is the quantized cinetic energy of the nuclei, and $W(x)$ represents the nuclei-nuclei interactions. Moreover, the parameter h is supposed to be small and, in the case of a molecular system, h^{-2} actually represents the quotient of electronic and nuclear masses. In more general systems, one can also include a magnetic potential and an exterior electric potential both in $\boldsymbol{\omega}$ and $Q(x)$. We refer to Chapter 12 for more details about this case.

We make the following assumptions:

$(H1)$ For all $\alpha, \beta \in Z\!\!Z_+^n$ with $|\alpha| \leq m$, $\partial^\beta c_\alpha(x,h) = \mathcal{O}(1)$ uniformly for $x \in I\!R^n$ and $h > 0$ small enough. Moreover, setting $\omega(x,\xi;h) := \sum_{|\alpha|\leq m} c_\alpha(x;h)\xi^\alpha$, we assume that there exists a constant $C_0 \geq 1$ such that, for all $(x,\xi) \in I\!R^{2n}$ and $h > 0$ small enough,

$$\mathrm{Re}\,\omega(x,\xi;h) \geq \frac{1}{C_0}\langle\xi\rangle^m - C_0.$$

In particular, Assumption $(H1)$ implies that m is even and $\boldsymbol{\omega}$ is well defined as a selfadjoint operator on $L^2(I\!R^n)$ (and, by extension, on $L^2(I\!R^n;\mathcal{H})$) with domain $H^m(I\!R^n)$. Moreover, by the Sharp Gårding Inequality (see, e.g., [**Ma2**]), it is uniformly semi-bounded from below.

$(H2)$ $W \geq 0$ is $\langle D_x\rangle^m$-compact on $L^2(I\!R^n)$, and there exists $\gamma \in I\!R$ such that, for all $x \in I\!R^n$, $Q(x) \geq \gamma$ on $\mathcal{H}$.

Assumptions $(H1)-(H2)$ guarantee that, for h sufficiently small, $P(h)$ can be realized as a selfadjoint operator on $L^2(\mathbb{R}^n;\mathcal{H})$ with domain $\mathcal{D}(P)\subset H^m(\mathbb{R}^n;\mathcal{H})\cap L^2(\mathbb{R}^n;\mathcal{D}_Q)$, and verifies $P(h)\geq \gamma_0$, with $\gamma_0\in\mathbb{R}$ independent of h.

(Of course, in the case of a molecular system, $P(h)$ is essentially selfadjoint, and the domain of its selfadjoint extension is $H^2(\mathbb{R}^n\times Y)$, where Y stands for the space of electron positions.)

For $L\geq 1$ and $L'\geq 0$, we denote by $\lambda_1(x)\leq\cdots\leq\lambda_{L+L'}(x)$ the first $L+L'$ values given by the Min-Max principle for $Q(x)$ on $\mathcal{H}$, and we make the following local gap assumption on the spectrum $\sigma(Q(x))$ of $Q(x)$:

$(H3)$ There exists a contractible bounded open set $\Omega\subset\mathbb{R}^n$ and $L\geq 1$ such that, for all $x\in\Omega$, $\lambda_1(x),\dots,\lambda_{L+L'}(x)$ are discrete eigenvalues of $Q(x)$, and one has,

$$\inf_{x\in\Omega}\text{ dist }(\sigma(Q(x))\backslash\{\lambda_{L'+1}(x),\dots,\lambda_{L'+L}(x)\},\{\lambda_{L'+1}(x),\dots,\lambda_{L'+L}(x)\})>0.$$

Furthermore, the spectral projections $\Pi_0^-(x)$, associated with $\{\lambda_1(x),\dots,\lambda_{L'}(x)\}$, and $\Pi_0(x)$, associated with $\{\lambda_{L'+1}(x),\dots,\lambda_{L'+L}(x)\}$, both depend continuously on $x\in\Omega$.

Then, we assume that P can be "regularized" with respect to x in Ω, in the following sense:

$(H4)$ There exists a finite family of bounded open sets $(\Omega_j)_{j=1}^r$ in $\mathbb{R}^n$, a corresponding family of unitary operators $U_j(x)$ ($j=1,\cdots,r$, $x\in\Omega_j$), and some fix selfadjoint operator $Q_0\geq C_0$ on $\mathcal{H}$ with domain $\mathcal{D}_Q$, such that (denoting by U_j the unitary operator on $L^2(\Omega_j;\mathcal{H})\simeq L^2(\Omega_j)\otimes\mathcal{H}$ induced by the action of $U_j(x)$ on $\mathcal{H}$),

- $\Omega=\cup_{j=1}^r\Omega_j$;
- For all $j=1,\cdots,r$ and $x\in\Omega_j$, $U_j(x)$ leaves $\mathcal{D}_Q$ invariant;
- For all j, the operator $U_j\boldsymbol{\omega}U_j^{-1}$ is a semiclassical differential operator with operator-valued symbol, of the form,

$$U_j\boldsymbol{\omega}U_j^{-1}=\boldsymbol{\omega}+h\sum_{|\beta|\leq m-1}\omega_{\beta,j}(x;h)(hD_x)^\beta, \tag{2.3}$$

  where $\omega_{\beta,j}Q_0^{\frac{|\beta|}{m}-1}\in C^\infty(\Omega_j;\mathcal{L}(\mathcal{H}))$ for any $\gamma\in\mathbb{Z}_+^n$ (here, $\mathcal{L}(\mathcal{H})$ stands for the Banach space of bounded operators on $\mathcal{H}$), and the quantity $\|\partial_x^\gamma\omega_{\beta,j}(x;h)Q_0^{\frac{|\beta|}{m}-1}\|_{\mathcal{L}(\mathcal{H})}$ is bounded uniformly with respect to h small enough and locally uniformly with respect to $x\in\Omega_j$;
- For all j, the operators $U_j(x)Q(x)U_j(x)^{-1}$ and $U_j(x)Q_0U_j(x)^{-1}$ are in $C^\infty(\Omega_j;\mathcal{L}(\mathcal{D}_Q,\mathcal{H}))$ (where $\mathcal{L}(\mathcal{D}_Q,\mathcal{H})$ stands for the space of bounded operators from $\mathcal{D}_Q$ to $\mathcal{H}$);
- $W\in C^\infty(\cup_{j=1}^r\Omega_j)$;
- There exists a dense subspace $\mathcal{H}_\infty\subset\mathcal{D}_Q\subset\mathcal{H}$, such that, for any $v\in\mathcal{H}_\infty$ and any $j=1,\cdots,r$, the application $x\mapsto U_j(x)v$ is in $C^\infty(\Omega_j,\mathcal{D}_Q)$.

Note that, for physical molecular systems, a construction of such operators $U_j(x)$'s is made in [**KMSW**], and can be performed around any point of $\mathbb{R}^n$ where W

is smooth. Moreover, in that case one can take $Q_0 = -\Delta_y + 1$ (where y stands for the position of the electrons), and the last point in (H4) can be realized by taking $\mathcal{H}_\infty = C_0^\infty(Y)$. Again, we refer the interested reader to Chapter 12. Let us also observe that, in the case $L' + L = 1$, one does not need to assume that Ω is contractible.

For any $\varphi_0 \in L^2(\mathbb{R}^n; \mathcal{H})$ (possibly h-dependent) such that $\|\varphi_0\|_{L^2(K_0^c;\mathcal{H})} = \mathcal{O}(h^\infty)$ for some compact set $K_0 \subset\subset \mathbb{R}^n$, and for any $\Omega' \subset\subset \mathbb{R}^n$ open neighborhood of K_0, we set,

$$T_{\Omega'}(\varphi_0) := \sup\{T > 0\,;\, \exists K_T \subset\subset \Omega',\ \sup_{t\in[0,T]} \|e^{-itP/h}\varphi_0\|_{L^2(K_T^c;\mathcal{H})} = \mathcal{O}(h^\infty)\}.$$

Then, $T_{\Omega'}(\varphi_0) \leq +\infty$, and, if one also assume that $\|(1 - f(P))\varphi_0\| = \mathcal{O}(h^\infty)$ for some $f \in C_0^\infty(\mathbb{R})$, Theorem B.1 in Appendix B shows that,

$$T_{\Omega'}(\varphi_0) \geq \frac{2 \text{ dist } (K_0, \partial\Omega')}{\|\nabla_\xi \omega(x, hD_x) g(P)\|},$$

for any $g \in C_0^\infty(\mathbb{R})$ verifying $gf = f$.

As a main result, we obtain (denoting by $L^2(\mathbb{R}^n)^{\oplus L}$ the space $(L^2(\mathbb{R}^n))^L$ endowed with its natural Hilbert structure),

THEOREM 2.1. *Assume (H1)-(H4) and let $\Omega' \subset\subset \Omega$ with Ω' open subset of $\mathbb{R}^n$. Then, for any $g \in C_0^\infty(\mathbb{R})$, there exists an orthogonal projection Π_g on $L^2(\mathbb{R}^n; \mathcal{H})$, an operator $\mathcal{W} : L^2(\mathbb{R}^n; \mathcal{H}) \to L^2(\mathbb{R}^n)^{\oplus L}$, uniformly bounded with respect to h, and a selfadjoint $L \times L$ matrix A of h-admissible operators $H^m(\mathbb{R}^n) \to L^2(\mathbb{R}^n)$, with the following properties:*

- *For all $\chi \in C_0^\infty(\Omega')$,*
$$\Pi_g \chi = \Pi_0 \chi + \mathcal{O}(h);$$
- *$\mathcal{W}\mathcal{W}^* = 1$ and $\mathcal{W}^*\mathcal{W} = \Pi_g$;*
- *For $x \in \Omega'$, the symbol $a(x, \xi; h)$ of A verifies,*
$$a(x, \xi; h) = \omega(x, \xi; h)\mathbf{I}_L + \mathcal{M}(x) + W(x)\mathbf{I}_L + hr(x, \xi; h)$$
where $\mathbf{I}_L$ stands for the L-dimensional identity matrix, $\mathcal{M}(x)$ is a $L \times L$ matrix depending smoothly on $x \in \Omega'$ and admitting $\lambda_{L'+1}(x)$, ..., $\lambda_{L'+L}(x)$ as eigenvalues, and where $\partial^\alpha r(x, \xi; h) = \mathcal{O}(\langle\xi\rangle^{m-1})$ for any multi-index α and uniformly with respect to $(x, \xi) \in \Omega' \times \mathbb{R}^n$ and $h > 0$ small enough;
- *For any $f \in C_0^\infty(\mathbb{R})$ with $\operatorname{Supp} f \subset \{g = 1\}$, and for any $\varphi_0 \in L^2(\mathbb{R}^n; \mathcal{H})$ such that $\|\varphi_0\| = 1$, and,*

$$\|\varphi_0\|_{L^2(K_0^c;\mathcal{H})} + \|(1 - \Pi_g)\varphi_0\| + \|(1 - f(P))\varphi_0\| = \mathcal{O}(h^\infty), \tag{2.4}$$

for some $K_0 \subset\subset \Omega'$, one has,

$$e^{-itP/h}\varphi_0 = \mathcal{W}^* e^{-itA/h}\mathcal{W}\varphi_0 + \mathcal{O}\left(\langle t\rangle h^\infty\right) \tag{2.5}$$

uniformly with respect to $h > 0$ small enough and $t \in [0, T_{\Omega'}(\varphi_0))$.

REMARK 2.2. *Actually, much more informations are obtained on the operators Π_g, $\mathcal{W}$ and A, and we refer to Theorems 7.1 and 8.1 for more details, and to Chapter 10 for an explicit computation of A, up to $\mathcal{O}(h^4)$.*

REMARK 2.3. *Condition (2.4) on the initial data may seems rather strong, but in fact, it will become clear from the proof that the operators* Π_g, $f(\tilde{P})$ *and* χ *(where* $\chi \in C_0^\infty(\mathbb{R}^n)$ *is supported in* K_0*) essentially commutes two by two (up to* $\mathcal{O}(h)$*). Indeed, in the case of a molecular system, they respectively correspond to a localization in space for the nuclei, a localization in energy for the electrons, and a localization in energy for the whole molecule.*

REMARK 2.4. *Here, we have assumed that both* $\Pi_0^-(x)$ *and* $\Pi_0(x)$ *have finite rank, since this corresponds to the main applications that we have in mind. However, it will become clear from the proof that the case where one or both of them have infinite rank could be treated in a similar way, with the difference that, if* $\mathrm{Rank}\,\Pi_0(x) = \infty$*, then* $\mathcal{W}^* e^{-itA/h}\mathcal{W}$ *must be replaced by* $e^{-it\Pi_g P \Pi_g/h}$ *(there will not be any operator* A *anymore). Moreover, some assumption must be added in order to be able to construct a modified operator as in Chapter 3 (for instance, that both* $\Pi_0^-(x)$ *and* $\Pi_0(x)$ *admit convenient extensions to all* $x \in \mathbb{R}^n$ *that depend smoothly on* x *away from a neighborhood of* K*).*

REMARK 2.5. *In the next chapter, we modify the operator* $Q(x)$ *away from the interesting region, in such a way that the new operator* $\tilde{Q}(x)$ *admits a* global *gap in its spectrum. With such an operator, a much better result can be obtained, that permits to decouple the evolution in a somewhat more complete and abstract way: see Theorem 7.1 (especially (7.2)). In particular, even if* $\|(1-\Pi_g)\varphi_0\|$ *is not small, Theorem 7.1 gives a description of the quantum evolution of* φ_0 *in terms of two independent reduced evolutions.*

As a corollary, in the case $L = 1$ we also obtain the following geometric lower bound on $T_{\Omega'}(\varphi_0)$, that relates it to the underlying classical Hamilton flow of the operator A:

COROLLARY 2.6. *Assume moreover that* $L = 1$ *and the coefficients* $c_\alpha = c_\alpha(x;h)$ *of* $\boldsymbol{\omega}$ *verify,*

$$c_\alpha(x;h) = c_{\alpha,0}(x) + \varepsilon(h)\tilde{c}_\alpha(x;h), \tag{2.6}$$

with $c_{\alpha,0}$ *real-valued and independent of* h, $\varepsilon(h) \to 0$ *as* $h \to 0$*, and, for any* β, $|\partial^\beta c_{\alpha,0}(x)| + |\partial^\beta \tilde{c}_\alpha(x,h)| = \mathcal{O}(1)$ *uniformly, and set,*

$$a_0(x,\xi) := \sum_{|\alpha|\le m} c_{\alpha,0}(x)\xi^\alpha + \lambda_{L'+1}(x) + W(x) \qquad (x \in \Omega').$$

Also, denote by $H_{a_0} := \partial_\xi a_0 \partial_x - \partial_x a_0 \partial_\xi$ *the Hamilton field of* a_0*. Then, for any* $f \in C_0^\infty(\mathbb{R})$ *with* $\mathrm{Supp}\, f \subset \{g = 1\}$*, and for any* $\varphi_0 \in L^2(\mathbb{R}^n;\mathcal{H})$ *such that* $\|\varphi_0\| = 1$*, and,*

$$\|\varphi_0\|_{L^2(K_0^c;\mathcal{H})} + \|(1-\Pi_g)\varphi_0\| + \|(1-f(P))\varphi_0\| = \mathcal{O}(h^\infty),$$

one has,

$$T_{\Omega'}(\varphi_0) \ge \sup\{T > 0\,;\, \pi_x(\cup_{t\in[0,T]} \exp tH_{a_0}(K(f))) \subset \Omega'\}, \tag{2.7}$$

where π_x *stands for the projection* $(x,\xi) \mapsto x$*, and* $K(f)$ *is the compact subset of* $\mathbb{R}^{2n}$ *defined by,*

$$K(f) := \{(x,\xi)\,;\, x \in K_0,\ \omega(x,\xi) + \gamma \le C_f\}$$

with $\gamma = \inf_{x\in\Omega'} \inf \sigma(Q(x))$ *and* $C_f := \mathrm{Max}|\,\mathrm{Supp}\, f|$*.*

REMARK 2.7. *Thanks to (H1) and (H2), it is easy to see that* $\exp tH_{a_0}(x,\xi)$ *is well defined for all* $(t,x,\xi) \in \mathbb{R} \times \mathbb{R}^{2n}$.

REMARK 2.8. *Actually, as it will be seen in the proof, in (2.7) one can replace the set* $K(f)$ *by* $\cup_{j=1}^{r} FS(U_j \Pi_g \varphi_0)$, *where* FS *stands for the Frequency Set of locally* L^2 *functions introduced in* [**GuSt**] *(we refer to Chapter 9 for more details).*

REMARK 2.9. *Our proof would permit to state a similar result in the case* $L > 1$, *but under the additional assumption that the set* $\{\lambda_{L'+1}(x), \dots, \lambda_{L'+L}(x)\}$ *can be written as* $\{E_1(x), \dots, E_{L''}(x)\}$, *where the (possibly degenerate) eigenvalues* $E_j(x)$ *are such that* $E_j(x) \neq E_{j'}(x)$ *for* $j \neq j'$ *and* $x \in \Omega$. *In the general case where crossings may occur, such a type of result relies on the microlocal propagation of the Frequency Set for solutions of semiclassical matrix evolution problems (for which not much is known, in general).*

REMARK 2.10. *The proof also provides a very explicit and somehow optimal bound on* $T_{\Omega'}(\varphi_0)$ *in the case where* φ_0 *is a coherent state with respect to the* x*-variables: see Theorem 11.3 and (11.8).*

CHAPTER 3

A Modified Operator

In this chapter, we consider an arbitrary compact subset $K \subset\subset \Omega$ and an open neighborhood $\Omega_K \subset\subset \Omega$ of K. We also denote by Ω_0 an open subset of $\mathbb{R}^n$, with closure disjoint from $\overline{\Omega_K}$, and such that $(\Omega_j)_{j=0}^r$ covers all of $\mathbb{R}^n$, and we set $U_0 := \mathbf{1}$. The purpose of this chapter is to modify $Q(x)$ for x outside a neighborhood of K, in order to make it regular with respect to x there, and to deal with a global gap instead of a local one.

Due to the contractibility of Ω, we know that there exist $L' + L$ continuous functions $u_1, \dots, u_{L'+L}$ in $C(\Omega; \mathcal{H})$, such that the families $(u_1(x), \dots, u_{L'}(x))$ and $(u_{L'+1}(x), \dots, u_{L'+L}(x))$ span $\operatorname{Ran}\Pi_0^-(x)$ and $\operatorname{Ran}\Pi_0(x)$ respectively, for all $x \in \Omega$ (see, e.g., [**KMSW**]).

Then, following Lemma 1.1 in [**KMSW**], we first prove,

LEMMA 3.1. *For all $x \in \mathbb{R}^n$, there exist $\tilde{u}_1(x), \dots, \tilde{u}_{L'+L}(x)$ in $\mathcal{D}_Q$, such that the family $(\tilde{u}_1(x), \dots, \tilde{u}_{L'+L}(x))$ is orthonormal in $\mathcal{H}$ for all $x \in \mathbb{R}^n$, the families $(\tilde{u}_1(x), \dots, \tilde{u}_{L'}(x))$ and $(\tilde{u}_{L'+1}(x), \dots, \tilde{u}_{L'+L}(x))$ span $\operatorname{Ran}\Pi_0^-(x)$ and $\operatorname{Ran}\Pi_0(x)$, respectively, when $x \in \Omega_K$, and, for all $j = 0, 1, \cdots, r$ and $k = 1, \dots, L' + L$,*

$$U_j(x)\tilde{u}_k(x) \in C^\infty(\Omega_j; \mathcal{D}_Q).$$

Proof – Let $\zeta_1, \zeta_2 \in C^\infty(\mathbb{R}^n; [0, 1])$, such that $\operatorname{Supp}\zeta_1 \subset \Omega_0^c$, $\zeta_1 = 1$ on Ω_K and $\zeta_1^2 + \zeta_2^2 = 1$ everywhere. Since $u_1(x), \dots, u_{L'+L}(x)$ depend continuously on x in Ω, for any $\varepsilon > 0$ one can find a finite number of points $x_1, \cdots, x_N \in \operatorname{Supp}\zeta_1$ and a partition of unity $\chi_1, \cdots, \chi_N \in C_0^\infty(\Omega)$ on $\operatorname{Supp}\zeta_1$, such that, for all $k = 1, \dots, L' + L$,

$$\sup_{x \in \operatorname{Supp}\zeta_1} \|u_k(x) - \sum_{\ell=1}^N \chi_\ell(x) u_k(x_\ell)\|_{\mathcal{H}} \le \varepsilon.$$

On the other hand, using the last assertion of (H4), for any (k, ℓ) one can find $v_{k,\ell}$ in $\mathcal{D}_Q$, such that, $\|v_{k,\ell} - u_k(x_\ell)\|_{\mathcal{H}} \le \varepsilon$ and $U_j(x)v_{k,\ell} \in C^\infty(\Omega_j, \mathcal{D}_Q)$ for all $j = 1, \dots, r$. Moreover, it follows from $(H3)$ and $(H4)$ that, for all $j = 1, \cdots, r$,

$$U_j(x)\Pi_0^-(x)U_j^*(x) \text{ and } U_j(x)\Pi_0(x)U_j^*(x) \in C^\infty(\Omega_j, \mathcal{L}(\mathcal{H}, \mathcal{D}_Q)).$$

Therefore, if we set,

$$v_k(x) := \Pi_0^-(x) \sum_{\ell=1}^N \chi_\ell(x) v_{k,\ell} \quad (k = 1, \dots, L');$$

$$v_k(x) := \Pi_0(x) \sum_{\ell=1}^N \chi_\ell(x) v_{k,\ell} \quad (k = L' + 1, \dots, L' + L),$$

and since $\sum_{\ell=1}^{N}\chi_\ell(x) = 1$ on Supp ζ_1, we obtain (also using that $\Pi_0^-(x)u_k(x) = u_k(x)$ for $k \leq L'$, and $\Pi_0(x)u_k(x) = u_k(x)$ for $k \geq L'+1$),

$$\sup_{x\in \text{ Supp } \zeta_1} \|u_k(x) - v_k(x)\|_{\mathcal{H}} \leq 2\varepsilon$$
$$U_j(x)v_k(x) \in C^\infty(\Omega_j, \mathcal{D}_Q) \quad (j = 1, \ldots, r).$$

In particular, by taking ε small enough, we see that the families $(v_1(x), \ldots, v_{L'}(x))$ and $(v_{L'+1}(x), \ldots, v_{L'+L}(x))$ span $\operatorname{Ran}\Pi_0^-(x)$ and $\operatorname{Ran}\Pi_0(x)$, respectively, for $x \in \operatorname{Supp}\zeta_1$. Moreover, by Gram-Schmidt, this families can also be assumed to be orthonormal.

Then, using again the last point of (H4), one can find an orthonormal family $w_1, \ldots, w_{L'+L} \in \mathcal{D}_Q$, such that $|\langle w_m, u_k(x_\ell)\rangle| \leq \varepsilon$ for all $1 \leq k,m \leq L'+L$, $1 \leq \ell \leq N$, and $U_j(x)w_m \in C^\infty(\Omega_j, \mathcal{D}_Q)$ $(j = 1, \ldots, r)$. Thus, setting,

$$\tilde{w}_k(x) := \zeta_1(x)v_k(x) + \zeta_2(x)w_k,$$

we see that, for all $k, k' \in \{1, \ldots, L'+L\}$,

$$\langle \tilde{w}_k(x), \tilde{w}_{k'}(x)\rangle_{\mathcal{H}} = \delta_{k,k'} + \mathcal{O}(\varepsilon).$$

As a consequence, taking $\varepsilon > 0$ sufficiently small and orthonormalizing the family $(\tilde{w}_1(x), \ldots, \tilde{w}_{L'+L}(x))$, we obtain a new family $(\tilde{u}_1(x), \ldots, \tilde{u}_{L'+L}(x))$ that verifies all the properties required in the lemma. •

Then, (with the usual convention $\sum_{k=1}^{L'} = 0$ if $L' = 0$) we set,

$$\tilde{\Pi}_0^-(x) = \sum_{k=1}^{L'} \langle \cdot, \tilde{u}_k(x)\rangle_{\mathcal{H}}\tilde{u}_k(x),$$
$$\tilde{\Pi}_0(x) = \sum_{k=L'+1}^{L'+L} \langle \cdot, \tilde{u}_k(x)\rangle_{\mathcal{H}}\tilde{u}_k(x)$$

so that $\tilde{\Pi}_0^-(x)$ and $\tilde{\Pi}_0(x)$ are orthogonal projections of rank L' and L respectively, are orthogonal each other, coincide with $\Pi_0^-(x)$ and $\Pi_0(x)$ for x in Ω_K, and verify,

$$U_j(x)\tilde{\Pi}_0^-(x)U_j(x)^* \text{ and } U_j(x)\tilde{\Pi}_0(x)U_j(x)^* \in C^\infty(\Omega_j, \mathcal{L}(\mathcal{H})), \tag{3.1}$$

for all $j = 0, 1, \cdots, r$.

Now, with the help of $\tilde{\Pi}_0^-(x), \tilde{\Pi}_0(x)$, we modify $Q(x)$ outside a neighborhood of K as follows.

PROPOSITION 3.2. *Let $\Omega_K' \subset\subset \Omega_K$ be an open neighborhood of K. Then, for all $x \in I\!R^n$, there exists a selfadjoint operator $\tilde{Q}(x)$ on $\mathcal{H}$, with domain $\mathcal{D}_Q$, and uniformly semi-bounded from below, such that,*

$$\tilde{Q}(x) = Q(x) \quad \text{if } x \in \Omega_K'; \tag{3.2}$$
$$[\tilde{Q}(x), \tilde{\Pi}_0^-(x)] = [\tilde{Q}(x), \tilde{\Pi}_0(x)] = 0 \quad \text{for all } x \in I\!R^n, \tag{3.3}$$

and the application $x \mapsto U_j(x)\tilde{Q}(x)U_j(x)^{-1}$ is in $C^\infty(\Omega_j; \mathcal{L}(\mathcal{D}_Q, \mathcal{H}))$ for all $j = 0, 1, \cdots, r$. Moreover, the bottom of the spectrum of $\tilde{Q}(x)$ consists in $L'+L$ eigenvalues $\tilde{\lambda}_1(x), \ldots, \tilde{\lambda}_{L'+L}(x)$, and $\tilde{Q}(x)$ admits a global gap in its spectrum, in the *sense that,*

$$\inf_{x\in \mathbb{R}^n} \text{dist}\,(\sigma(\tilde{Q}(x))\backslash\{\tilde{\lambda}_{L'+1}(x),\dots,\tilde{\lambda}_{L'+L}(x)\},\{\tilde{\lambda}_{L'+1}(x),\dots,\tilde{\lambda}_{L'+L}(x)\})>0.$$

Proof We set $\tilde{\Pi}_0^+(x) = 1-\tilde{\Pi}_0^-(x)-\tilde{\Pi}_0(x)$ and we choose a function $\zeta \in C_0^\infty(\Omega_K;[0,1])$ such that $\zeta=1$ on Ω'_K. Then, with Q_0 as in (H4), we set,

$$\tilde{Q}(x)=\zeta(x)Q(x)+(1-\zeta(x))\tilde{\Pi}_0^+(x)Q_0\tilde{\Pi}_0^+(x)-(1-\zeta(x))\tilde{\Pi}_0^-(x).$$

Since $\tilde{\Pi}_0^-(x)=\Pi_0^-(x)$ and $\tilde{\Pi}_0(x)=\Pi_0(x)$ on $\text{Supp}\,\zeta$, we see that $\tilde{\Pi}_0^-(x)$ and $\tilde{\Pi}_0(x)$ commute with $\tilde{Q}(x)$, and it is also clear that $\tilde{Q}(x)$ is selfadjoint with domain $\mathcal{D}_Q$. Moreover,

$$\begin{aligned}&\tilde{\Pi}_0^-(x)\tilde{Q}(x)\tilde{\Pi}_0^-(x)=\zeta(x)\Pi_0^-(x)Q(x)\Pi_0^-(x)-(1-\zeta(x))\Pi_0^-(x);\\&\tilde{\Pi}_0(x)\tilde{Q}(x)\tilde{\Pi}_0(x)=\zeta(x)\Pi_0(x)Q(x)\Pi_0(x),\end{aligned}$$

and, setting,

$$\lambda_{L+L'+1}(x):=\inf\left(\sigma(Q(x))\backslash\{\lambda_1(x),\dots,\lambda_{L+L'}(x)\}\right),$$

one has,

$$\tilde{\Pi}_0^+(x)\tilde{Q}(x)\tilde{\Pi}_0^+(x)\geq(\zeta(x)\lambda_{L+L'+1}(x)+(1-\zeta(x))\,\tilde{\Pi}_0^+(x).$$

In particular, the bottom of the spectrum of $\tilde{Q}(x)$ consists in the $L+L'$ eigenvalues $\tilde{\lambda}_k(x)=\zeta(x)\lambda_k(x)-(1-\zeta(x))$ $(k=1,\dots,L')$, $\tilde{\lambda}_k(x)=\zeta(x)\lambda_k(x)$ $(k=L'+1,\dots,L'+L)$, and, due to (H3), one has,

$$\inf_{x\in\mathbb{R}^n}\left(\tilde{\lambda}_{L'+1}(x)-\tilde{\lambda}_{L'}(x)\right)=\inf_{x\in\mathbb{R}^n}\left(\zeta(x)(\lambda_{L'+1}(x)-\lambda_{L'}(x)+(1-\zeta(x))\right)>0,$$

and

$$\begin{aligned}&\inf_{x\in\Omega}\text{dist}\,(\sigma(\tilde{Q}(x))\backslash\{\tilde{\lambda}_1(x),\dots,\tilde{\lambda}_{L'+L}(x)\},\{\tilde{\lambda}_1(x),\dots,\tilde{\lambda}_{L'+L}(x)\})\\&\qquad\qquad\geq\inf_{x\in\Omega}|\zeta(x)(\lambda_{L'+L+1}(x)-\lambda_{L'+L}(x))+(1-\zeta(x))|>0,\end{aligned}$$

while, since $\text{Supp}\,\zeta\subset\Omega$,

$$\inf_{x\in\mathbb{R}^n\backslash\Omega}\text{dist}\,(\sigma(\tilde{Q}(x))\backslash\{\tilde{\lambda}_1(x),\dots,\tilde{\lambda}_{L'+L}(x)\},\{\tilde{\lambda}_1(x),\dots,\tilde{\lambda}_{L'+L}(x)\})\geq 1.$$

In particular, $\tilde{Q}(x)$ admits a fix global gap in its spectrum as stated in the proposition. Finally, using (H4) and (3.1), we see that $U_j(x)\tilde{Q}(x)U_j^*(x)$ depends smoothly on x in Ω_j for all $j=0,1,\cdots,r$. •

In the sequel, we also set,

$$\tilde{P}=\boldsymbol{\omega}+\boldsymbol{Q}:=\boldsymbol{\omega}+\tilde{Q}(x)+\zeta(x)W(x), \tag{3.4}$$

and we denote by $\tilde{\Pi}_0$ the projection on $L^2(\mathbb{R}^n;\mathcal{H})$ induced by the action of $\tilde{\Pi}_0(x)$ on $\mathcal{H}$, i.e. the unique projection on $L^2(\mathbb{R}^n;\mathcal{H})$ that verifies

$$\tilde{\Pi}_0(f\otimes g)(x)=f(x)\tilde{\Pi}_0(x)g\quad(\text{a.e. on }\mathbb{R}^n\ni x)$$

for all $f\in L^2(\mathbb{R}^n)$ and $g\in\mathcal{H}$.

CHAPTER 4

Twisted h-Admissible Operators

In order to construct (in the same spirit as in [**BrNo, HeSj12, MaSo, NeSo, Sj2, So**]) an orthogonal projection Π on $L^2(I\!R^n;\mathcal{H})$ such that $\Pi - \Pi_0 = \mathcal{O}(h)$ and $[\tilde{P},\Pi] = \mathcal{O}(h^\infty)$ (locally uniformly in energy), we need to generalize the notion of h-admissible operator with operator-valued symbol (see, e.g., [**Ba, GMS**] and the Appendix) by taking into account the possible singularities of $Q(x)$. To avoid complications, in this chapter we also restrict our attention to the case of bounded operators. The case of unbounded ones will be considered in the next chapter, at least from the point of view of *differential* operators.

DEFINITION 4.1. *We call "regular covering" of $I\!R^n$ any finite family $(\Omega_j)_{j=0,\cdots,r}$ of open subsets of $I\!R^n$ such that $\cup_{j=0}^r\Omega_j = I\!R^n$ and such that there exists a family of functions $\chi_j \in C_b^\infty(I\!R^n)$ (the space of smooth functions on $I\!R^n$ with uniformly bounded derivatives of all order) with $\sum_{j=0}^r \chi_j = 1$, $0 \leq \chi_j \leq 1$, and* $\operatorname{dist}(\operatorname{Supp}(\chi_j), I\!R^n\backslash\Omega_j) > 0$ $(j = 0,\cdots,r)$. *Moreover, if $U_j(x)$ $(x\in\Omega_j,\ 0\leq j\leq r)$ is a family of unitary operators on $\mathcal{H}$, the family $(U_j,\Omega_j)_{j=0,\cdots,r}$ (where U_j denotes the unitary operator on $L^2(\Omega_j;\mathcal{H}) \simeq L^2(\Omega_j)\otimes\mathcal{H}$ induced by the action of $U_j(x)$ on $\mathcal{H}$) will be called a "regular unitary covering" of $L^2(I\!R^n;\mathcal{H})$.*

REMARK 4.2. *Despite the terminology that we use, no assumption is made on any possible regularity of $U_j(x)$ with respect to x.*

REMARK 4.3. *Possibly by shrinking a little bit Ω around the compact set K, one can always assume that the family $(U_j,\Omega_j)_{j=0,1,\cdots,r}$ defined in Chapter 2 is a regular unitary covering of $L^2(I\!R^n;\mathcal{H})$.*

In the sequels, we denote by $C_d^\infty(\Omega_j)$ the space of functions $\chi \in C_b^\infty(I\!R^n)$ such that $\operatorname{dist}(\operatorname{Supp}(\chi), I\!R^n\backslash\Omega_j) > 0$.

DEFINITION 4.4 (Twisted h-Admissible Operator). *Let $\mathcal{U} := (U_j,\Omega_j)_{j=0,\cdots,r}$ be a regular unitary covering (in the previous sense) of $L^2(I\!R^n;\mathcal{H})$. We say that an operator $A : L^2(I\!R^n;\mathcal{H}) \to L^2(I\!R^n;\mathcal{H})$ is a $\mathcal{U}$-twisted h-admissible operator, if there exists a family of functions $\chi_j \in C_d^\infty(\Omega_j)$ such that, for any $N\geq 1$, A can be written in the form,*

$$A = \sum_{j=0}^{r} U_j^{-1}\chi_j A_j^N U_j\chi_j + R_N, \tag{4.1}$$

where $\|R_N\|_{\mathcal{L}(L^2(I\!R^n;\mathcal{H}))} = \mathcal{O}(h^N)$, and, for any $j = 0,..,r$, A_j^N is a bounded h-admissible operator on $L^2(I\!R^n;\mathcal{H})$ with symbol $a_j^N(x,\xi)\in C_b^\infty(T^*I\!R^n;\mathcal{L}(\mathcal{H}))$, and, for any $\varphi_\ell \in C_d^\infty(\Omega_\ell)$ $(\ell = 0,..,r)$, the operator

$$U_\ell\varphi_\ell U_j^{-1}\chi_j A_j^N \chi_j U_j U_\ell^{-1}\varphi_\ell,$$

is still an h-admissible operator on $L^2(\mathbb{R}^n;\mathcal{H})$.

REMARK 4.5. *In particular, by the Calderón-Vaillancourt theorem, the norm of A on $L^2(\mathbb{R}^n;\mathcal{H})$ is bounded uniformly with respect to $h\in(0,1]$.*

An equivalent definition is given by the following proposition:

PROPOSITION 4.6. *An operator $A : L^2(\mathbb{R}^n;\mathcal{H}) \to L^2(\mathbb{R}^n;\mathcal{H})$ is a $\mathcal{U}$-twisted h-admissible operator if and only if the two following properties are verified:*

(1) *For any $N\geq 1$ and any functions $\chi_1,\cdots,\chi_N\in C_b^\infty(\mathbb{R}^n)$, one has,*

$$\mathrm{ad}_{\chi_1}\circ\cdots\circ\mathrm{ad}_{\chi_N}(A)=\mathcal{O}(h^N)\quad :\ L^2(\mathbb{R}^n;\mathcal{H})\to L^2(\mathbb{R}^n;\mathcal{H})$$

where we have used the notation $\mathrm{ad}_\chi(A):=[\chi,A]=\chi A-A\chi$.

(2) *For any $\varphi_j\in C_d^\infty(\Omega_j)$, $U_j\varphi_jAU_j^{-1}\varphi_j$ is a bounded h-admissible operator on $L^2(\mathbb{R}^n;\mathcal{H})$.*

Proof – From Definition 4.4, it is clear that any $\mathcal{U}$-twisted h-admissible operator verifies the properties of the Proposition. Conversely, assume A verifies these properties, and denote by $(\chi_j)_{j=0,\cdots,r}\subset C_b^\infty(\mathbb{R}^n)$ a partition of unity on $\mathbb{R}^n$ such that dist (Supp $(\chi_j),\mathbb{R}^n\backslash\Omega_j) > 0$. Then, for all j one can construct $\varphi_j,\psi_j\in C_d^\infty(\Omega_j)$, such that $\varphi_j\chi_j=\chi_j$ and $\psi_j\varphi_j=\varphi_j$, and, for any $N\geq 1$, we can write,

$$\begin{aligned}
A &= \sum_{j=0}^r\chi_jA=\sum_{j=0}^r\left(\chi_jA\varphi_j+\chi_j\mathrm{ad}_{\varphi_j}(A)\right)\\
&= \sum_{j=0}^r\left(\chi_jA\varphi_j+\chi_j\mathrm{ad}_{\varphi_j}(A)\varphi_j+\chi_j\mathrm{ad}_{\varphi_j}^2(A)\right)\\
&= \cdots=\sum_{j=0}^r\left(\sum_{k=0}^{N-1}\chi_j\mathrm{ad}_{\varphi_j}^k(A)\varphi_j+\chi_j\mathrm{ad}_{\varphi_j}^N(A)\right)\\
&= \sum_{j=0}^r\left(\sum_{k=0}^{N-1}\psi_j\chi_j\mathrm{ad}_{\varphi_j}^k(A)\varphi_j\psi_j+\chi_j\mathrm{ad}_{\varphi_j}^N(A)\right).
\end{aligned}$$

In particular, since $\mathrm{ad}_{\varphi_j}^N(A)=\mathcal{O}(h^N)$, and U_j commutes with the multiplication by functions of x, we obtain

$$A=\sum_{j=0}^r U_j^{-1}\psi_jA_j^NU_j\psi_j+\mathcal{O}(h^N)\tag{4.2}$$

with

$$A_j^N:=\sum_{k=0}^{N-1}U_j\chi_j\mathrm{ad}_{\varphi_j}^k(A)U_j^{-1}\varphi_j=\sum_{k=0}^{N-1}\chi_j\mathrm{ad}_{\varphi_j}^k(U_j\varphi_jAU_j^{-1}\varphi_j).\tag{4.3}$$

Therefore, A_j^N is a bounded h-admissible operator, and for any $\tilde{\psi}_l\in C_d^\infty(\Omega_l)$, it verifies,

$$U_l\tilde{\psi}_lU_j^{-1}\psi_jA_j^N\psi_jU_j\tilde{\psi}_lU_l^{-1}=\sum_{k=0}^{N-1}\chi_j\mathrm{ad}_{\varphi_j}^k(U_l\tilde{\psi}_lAU_l^{-1}\tilde{\psi}_l)\varphi_j,$$

that is still an h-admissible operator. Thus, the proposition follows. •

In the sequel, if A is a $\mathcal{U}$-twisted h-admissible operator, then an expression of A as in (4.1) will be said "adapted" to $\mathcal{U}$.

One also has at disposal a notion of (full) symbol for such operators. In the sequels, we denote by $S(\Omega_j \times \mathbb{R}^n; \mathcal{L}(\mathcal{H}))$ the space of (h-dependent) operator-valued symbols $a_j \in C^\infty(\Omega_j \times \mathbb{R}^n; \mathcal{L}(\mathcal{H}))$ such that, for any $\alpha \in \mathbb{Z}_+^{2n}$, the quantity $\|\partial^\alpha a_j(x,\xi)\|_{\mathcal{L}(\mathcal{H})}$ is bounded uniformly for h small enough and for (x,ξ) in any set of the form $\Omega_j' \times \mathbb{R}^n$, with $\Omega_j' \subset \Omega_j$, $\text{dist}\left(\Omega_j', \mathbb{R}^n \backslash \Omega_j\right) > 0$. We also set,

$$\begin{aligned}
&\Omega := (\Omega_j)_{j=0,\dots,r};\\
&\mathbf{S}(\Omega; \mathcal{L}(\mathcal{H})) := S(\Omega_0 \times \mathbb{R}^n; \mathcal{L}(\mathcal{H})) \times \cdots \times S(\Omega_r \times \mathbb{R}^n; \mathcal{L}(\mathcal{H})),
\end{aligned}$$

and we write $a = \mathcal{O}(h^\infty)$ in $\mathbf{S}(\Omega; \mathcal{L}(\mathcal{H}))$ when $\|\partial^\alpha a_j(x,\xi)\|_{\mathcal{L}(\mathcal{H})} = \mathcal{O}(h^\infty)$ uniformly in any set $\Omega_j' \times \mathbb{R}^n$ as before.

LEMMA 4.7. *Let A be a $\mathcal{U}$-twisted h-admissible operator, where $\mathcal{U} = (U_j, \Omega_j)_{0\le j\le r}$ is some regular unitary covering. Then, for all $j = 0, \dots, r$, there exists an operator-valued symbol $a_j \in S(\Omega_j \times \mathbb{R}^n; \mathcal{L}(\mathcal{H}))$, unique up to $\mathcal{O}(h^\infty)$, such that, for any $\chi_j = \chi_j(x) \in C_d^\infty(\Omega_j)$, the symbol of the h-admissible operator $U_j\chi_j A U_j^{-1}\chi_j$ is $\chi_j \sharp a_j \sharp \chi_j$ (where $\sharp$ stands for the standard symbolic composition: see Appendix A).*

Proof – Indeed, given two functions $\chi_j, \varphi_j \in C_d^\infty(\Omega_j)$ with $\varphi_j\chi_j = \chi_j$, one has

$$U_j\chi_j A U_j^{-1}\chi_j = \chi_j \left(U_j\varphi_j A U_j^{-1}\varphi_j\right)\chi_j,$$

and thus, denoting by a_j^χ the symbol of $U_j\chi A U_j^{-1}\chi$, one obtains

$$a_j^{\chi_j} = \chi_j \sharp a_j^{\varphi_j} \sharp \chi_j.$$

In particular, using the explicit expression of $\sharp$ (see Appendix A, Proposition A.2), we see that $a_j^{\varphi_j} = a_j^{\chi_j} + \mathcal{O}(h^\infty)$ in the interior of $\{\chi_j(x) = 1\}$. Then, the result follows by taking a non-decreasing sequence $(\varphi_{j,k})_{k\ge1}$ in $C_d^\infty(\Omega_j)$, such that $\bigcup_{k\ge0}\{x \in \Omega_j\,;\, \varphi_{j,k}(x) = 1\} = \Omega_j$, and, for any $(x,\xi) \in \Omega_j \times \mathbb{R}^n$, by defining $a_j(x,\xi)$ as the common value of the $a_j^{\varphi_{j,k}}(x,\xi)$'s for k large enough. •

DEFINITION 4.8 (Symbol). *Let A be a $\mathcal{U}$-twisted h-admissible operator, where $\mathcal{U} = (U_j, \Omega_j)_{0\le j\le r}$ is some regular unitary covering. Then, the family of operator-valued functions $\sigma(A) := (a_j)_{0\le j\le r} \in \mathbf{S}(\Omega; \mathcal{L}(\mathcal{H}))$, defined in the previous lemma, is called the (full) symbol of A. Moreover, A is said to be elliptic if, for any $j = 0, \cdots, r$ and $(x,\xi) \in \Omega_j \times \mathbb{R}^n$, the operator $a_j(x,\xi)$ is invertible on $\mathcal{H}$, and verifies,*

$$\|a_j(x,\xi)^{-1}\|_{\mathcal{L}(\mathcal{H})} = \mathcal{O}(1), \tag{4.4}$$

uniformly for h small enough and for (x,ξ) in any set of the form $\Omega_j' \times \mathbb{R}^n$, with $\Omega_j' \subset \Omega_j$, $\text{dist}\left(\Omega_j', \mathbb{R}^n \backslash \Omega_j\right) > 0$.

In particular, it follows from the proof of Proposition 4.6 that, if such an operator A is elliptic, then it can be written in the form (4.1), with A_j^N elliptic on $\{\chi_j \neq 0\}$ for all j, N. Moreover, we have the two following result on composition and parametrices:

PROPOSITION 4.9 (Composition). *Let $\mathcal{U}$ be a regular covering of $L^2(\mathbb{R}^n;\mathcal{H})$, and let A and B be two $\mathcal{U}$-twisted h-admissible operators. Then, the composition AB is a $\mathcal{U}$-twisted h-admissible operator, too. Moreover, its symbol is given by,*

$$\sigma(AB) = \sigma(A)\sharp\sigma(B),$$

where the operation $\sharp$ is defined component by component, that is,

$$(a_j)_{0\leq j\leq r}\sharp(b_j)_{0\leq j\leq r} := (a_j\sharp b_j)_{0\leq j\leq r}.$$

Proof – First of all, since

$$\mathrm{ad}_{\chi}(AB) = \mathrm{ad}_{\chi}(A)B + A\,\mathrm{ad}_{\chi}(B),$$

one easily sees, by induction on N, that the first condition in Proposition 4.6 is satisfied. Moreover, if $\chi_j \in C_d^\infty(\Omega_j)$, let $\varphi_j \in C_d^\infty(\Omega_j)$ such that $\varphi_j\chi_j = \chi_j$. Then, if, for any operator C, we set $C_j := U_j\varphi_j C U_j^{-1}\varphi_j$, we have,

$$\begin{aligned}
U_j\chi_j ABU_j^{-1}\chi_j &= \chi_j A_j B_j\chi_j + U_j\chi_j\mathrm{ad}_{(\varphi_j^2)}(A)BU_j^{-1}\chi_j\\
&= \chi_j A_j B_j\chi_j + \chi_j[\mathrm{ad}_{(\varphi_j^2)}(A)]_j B_j\chi_j + U_j\chi_j\mathrm{ad}^2_{(\varphi_j^2)}(A)BU_j^{-1}\chi_j\\
&= \cdots\\
&= \sum_{k=0}^{N-1}\chi_j[\mathrm{ad}^k_{(\varphi_j^2)}(A)]_j B_j\chi_j + U_j\chi_j\mathrm{ad}^N_{(\varphi_j^2)}(A)BU_j^{-1}\chi_j
\end{aligned}\tag{4.5}$$

for all $N\geq 1$. Therefore, since $U_j\chi_j\mathrm{ad}^N_{(\varphi_j^2)}(A)BU_j^{-1}\chi_j = \mathcal{O}(h^N)$, and the operator $[\mathrm{ad}^k_{(\varphi_j^2)}(A)]_j = \mathrm{ad}^k_{(\varphi_j^2)}(A_j)$ is a bounded h-admissible operator, we deduce from (4.5) that AB is a $\mathcal{U}$-twisted h-admissible operator. Moreover, since $\varphi_j = 1$ on the support of χ_j, we see that the symbol of $\chi_j\mathrm{ad}^k_{(\varphi_j^2)}(A_j)$ vanishes identically for $k\geq 1$, and thus, we also deduce from (4.5) that the symbol $(c_j)_{0\leq j\leq r}$ of AB verifies,

$$\chi_j\sharp c_j\sharp\chi_j = \chi_j\sharp a_j\sharp b_j\sharp\chi_j,$$

for any $\chi_j \in C_d^\infty(\Omega_j)$, and the result follows. •

PROPOSITION 4.10 (Parametrix). *Let A be a $\mathcal{U}$-twisted h-admissible operator, and assume that A is elliptic. Then, A is invertible on $L^2(\mathbb{R}^n;\mathcal{H})$, and its inverse A^{-1} is a $\mathcal{U}$-twisted h-admissible operator. Moreover, its symbol $\sigma(A^{-1})$ is related to the one $\sigma(A) = (a_j)_{0\leq j\leq r}$ of A by the following formula:*

$$\sigma(A^{-1}) = (\sigma(A))^{-1} + hb,$$

where $(\sigma(A))^{-1} := (a_j^{-1})_{0\leq j\leq r}$ and $b\in\mathbf{S}(\Omega;\mathcal{L}(\mathcal{H}))$.

Proof – We first prove that A is invertible by following an idea of [**KMSW**] (proof of Theorem 1.2).

For $j = 0,\cdots,r$, let $\chi_j,\varphi_j \in C_d^\infty(\Omega_j)$ such that $\varphi_j\chi_j = \chi_j$, and $\sum_{j=0}^r\chi_j = 1$. Then, by assumption, the symbol of $U_j\varphi_j AU_j^{-1}\varphi_j$ can be written on the form $\varphi_j(x)\sharp a_j(x,\xi)\sharp\varphi_j(x)$ with $a_j(x,\xi)$ invertible, and the operator,

$$B := \sum_{j=0}^r U_j^{-1}\varphi_j^3\mathrm{Op}_h(\varphi_j a_j^{-1})U_j\chi_j$$

is well defined and bounded on $L^2(I\!R^n;\mathcal{H})$. Moreover, using the standard symbolic calculus, we compute,

$$\begin{aligned} AB &= \sum_{j=0}^{r} AU_j^{-1}\varphi_j^3\mathrm{Op}_h(\varphi_j a_j^{-1})U_j\chi_j \\ &= \sum_{j=0}^{r} U_j^{-1}\varphi_j U_j\varphi_j AU_j^{-1}\varphi_j\mathrm{Op}_h(\varphi_j a_j^{-1})U_j\chi_j \\ &\qquad\qquad +[A,\varphi_j^2]U_j^{-1}\varphi_j\mathrm{Op}_h(\varphi_j a_j^{-1})U_j\chi_j \\ &= \sum_{j=0}^{r} U_j^{-1}\varphi_j\mathrm{Op}_h(\varphi_j^2 a_j)\mathrm{Op}_h(\varphi_j a_j^{-1})U_j\chi_j + \mathcal{O}(h) \\ &= \sum_{j=0}^{r} U_j^{-1}\varphi_j^4 U_j\chi_j + \mathcal{O}(h) = \sum_{j=0}^{r}\chi_j + \mathcal{O}(h) = 1+\mathcal{O}(h). \qquad (4.6) \end{aligned}$$

In the same way, defining,

$$B' := \sum_{j=0}^{r} U_j^{-1}\chi_j\mathrm{Op}_h(\varphi_j a_j^{-1})U_j\varphi_j^3,$$

we obtain $B'A = 1+\mathcal{O}(h)$, and this proves the invertibility of A for h small enough. It remains to verify that A^{-1} is a $\mathcal{U}$-twisted h-admissible operator. We first prove,

LEMMA 4.11. *Let A be a $\mathcal{U}$-twisted h-admissible operator, and let $\chi,\psi \in C_b^\infty(I\!R^n)$ such that* $\operatorname{dist}(\operatorname{Supp}\chi, \operatorname{Supp}\psi) > 0$. *Then,* $\|\chi A\psi\| = \mathcal{O}(h^\infty)$.

Proof – Given $N \geq 1$, let $\varphi_1,\cdots,\varphi_N \in C_b^\infty(I\!R^n)$, such that $\varphi_1\chi = \chi$, $\varphi_{k+1}\varphi_k = \varphi_k$ $(k=1,\cdots,N-1)$, and $\varphi_N\psi = 0$. Then, one has,

$$\begin{aligned} \chi A\psi &= \varphi_1\mathrm{ad}_\chi(A)\psi = \varphi_2\mathrm{ad}_{\varphi_1}\circ\mathrm{ad}_\chi(A)\psi \\ &= \cdots = \mathrm{ad}_{\varphi_N}\circ\cdots\circ\mathrm{ad}_{\varphi_1}\circ\mathrm{ad}_\chi(A)\psi = \mathcal{O}(h^{N+1}). \end{aligned}$$

•

Now, since,

$$\mathrm{ad}_\chi(A^{-1}) = -A^{-1}\mathrm{ad}_\chi(A)A^{-1},$$

it is easy to see, by induction on N, that A^{-1} satisfies to the first property of Proposition 4.6. Moreover, for $v \in L^2(I\!R^n;\mathcal{H})$ and for $\chi_j \in C_d^\infty(\Omega_j)$, let us set,

$$u = A^{-1}U_j^{-1}\chi_j v,$$

and choose $\varphi_j, \in C_d^\infty(\Omega_j;I\!R)$, $\psi_j \in C_b^\infty(I\!R^n;I\!R)$, such that $\psi_j\chi_j = 0$, $\varphi_j^4+\psi_j^2 \geq 1$, and $\operatorname{dist}(\operatorname{Supp}(\varphi_j-1), \operatorname{Supp}\chi_j) > 0$. Then, since the symbol of $A_j := U_j\varphi_j AU_j^{-1}\varphi_j$ is of the form $\varphi_j\sharp a_j\sharp\varphi_j$ with $a_j(x,\xi)$ invertible for x in $\operatorname{Supp}\varphi_j$, we see that the bounded h-admissible operator $B_j := A_j^*A_j+\psi_j^2$ is globally elliptic, and one has,

$$\begin{aligned} B_jU_j\chi_j u &= A_j^*A_jU_j\chi_j u = A_j^*U_j\varphi_j A\chi_j u = A_j^*U_j\chi_j Au + A_j^*U_j\varphi_j[A,\chi_j]u \\ &= A_j^*\chi_j^2 v + A_j^*U_j\varphi_j[A,\chi_j]\varphi_j^2 u + A_j^*U_j\chi_j A(\varphi_j^2-1)u \\ &= A_j^*\chi_j^2 v + A_j^*[A_j,\chi_j]U_j\varphi_j u + \mathcal{O}(h^\infty\|v\|), \qquad (4.7) \end{aligned}$$

where the last estimate comes from Lemma 4.11. In particular, since B_j^{-1} is an h-admissible operator, we obtain that $U_j\chi_j u$ can be written on the form,

$$U_j\chi_j u = C_j v + hC_j' U_j\varphi_j u + \mathcal{O}(h^\infty \|v\|)$$

where C_j, C_j' are bounded h-admissible operators. Repeating the same argument with $U_j\varphi_j u$ instead of $U_j\chi_j u$, and iterating the procedure, it easily follows that $U_j\chi_j A^{-1}U_j^{-1}\chi_j$ is an h-admissible operator. Moreover, we see on (4.7) that the symbol of $U_j\chi_j A^{-1}U_j^{-1}\chi_j$ coincides, up to $\mathcal{O}(h)$, with that of $B_j^{-1}A_j^*\chi_j^2$, that is,

$$(\varphi_j^4(x)a_j^*(x,\xi)a_j(x,\xi) + \psi_j^2(x))^{-1}a_j^*(x,\xi)\chi_j(x)^2 = a_j(x,\xi)^{-1}\chi_j(x)^2,$$

since $\varphi_j = 1$ and $\psi_j = 0$ on the support of χ_j. Thus, the proposition follows. •

PROPOSITION 4.12 (Functional Calculus). *Let A be a selfadjoint $\mathcal{U}$-twisted h-admissible operator, and let $f \in C_0^\infty(\mathbb{R})$. Then, the operator $f(A)$ is a $\mathcal{U}$-twisted h-admissible operator, and its symbol is related to that of A by the formula,*

$$\sigma(f(A)) = f(\operatorname{Re}\sigma(A)) + hb,$$

where $f(\operatorname{Re}(a_j)_{j=0,\dots,r}) := (f(\operatorname{Re} a_j))_{j=0,\dots,r}$, $\operatorname{Re} a_j := \frac{1}{2}(a_j + a_j^*)$, *and* $b \in \mathbf{S}(\Omega; \mathcal{L}(H))$.

Proof – We use a formula of representation of $f(A)$ due to B. Helffer and J. Sjöstrand. Denote by $\tilde{f} \in C_0^\infty(\mathbb{C})$ an almost analytic extension of f, that is, a function verifying $\tilde{f}|_{\mathbb{R}} = f$ and $|\overline{\partial}\tilde{f}(z)| = \mathcal{O}(|\operatorname{Im} z|^\infty)$ uniformly on $\mathbb{C}$. Then, we have (see, e.g., [**DiSj1, Ma2**]),

$$f(A) = \frac{1}{\pi}\int_{\mathbb{C}} \overline{\partial}\tilde{f}(z)(A-z)^{-1} d\operatorname{Re} z\, d\operatorname{Im} z. \tag{4.8}$$

Now, by Proposition 4.10, we see that, for $z \in \mathbb{C}\backslash\mathbb{R}$, the operator $(A-z)^{-1}$ is a $\mathcal{U}$-twisted h-admissible operator. Moreover, by standard rules on the operations ad_χ, if A and B are two bounded operators, then, for any $N \geq 1$ and any $\chi_1, \cdots, \chi_N \in C_b^\infty(\mathbb{R}^n)$, one has,

$$\mathrm{ad}_{\chi_1} \circ \cdots \circ \mathrm{ad}_{\chi_N}(AB) = \sum_{\substack{I\cup J=\{1,\dots,N\}\\ I\cap J=\emptyset}} \left(\prod_{i\in I}\mathrm{ad}_{\chi_i}\right)(A)\left(\prod_{j\in J}\mathrm{ad}_{\chi_j}\right)(B).$$

In particular, replacing A and B by $A-z$ and $(A-z)^{-1}$ respectively, one obtains,

$$\begin{aligned}&\mathrm{ad}_{\chi_1} \circ \cdots \circ \mathrm{ad}_{\chi_N}((A-z)^{-1})\\ &= -(A-z)^{-1}\sum_{\substack{I\cup J=\{1,\dots,N\}\\ I\cap J=\emptyset,\ I\neq\emptyset}} \left(\prod_{i\in I}\mathrm{ad}_{\chi_i}\right)(A-z)\left(\prod_{j\in J}\mathrm{ad}_{\chi_j}\right)((A-z)^{-1}),\end{aligned}$$

and thus, an easy induction gives,

$$\mathrm{ad}_{\chi_1} \circ \cdots \circ \mathrm{ad}_{\chi_N}((A-z)^{-1}) = \mathcal{O}(h^N|\operatorname{Im} z|^{-(N+1)}),$$

uniformly with respect to h and z. Therefore, it is immediate from (4.8) that $f(A)$ verifies the first condition in Proposition 4.6.

Moreover, setting $(a_j)_{0\leq j\leq r} := \sigma(A)$, for $\chi_j \in C_d^\infty(\omega_j)$, we denote by $B_j(z)$ the h-admissible operator with symbol $(\operatorname{Re} a_j - \overline{z})(\operatorname{Re} a_j - z)\varphi_j^4 + \psi_j^2$, where φ_j

and ψ_j are as at the end of the proof of Proposition 4.10. Then, using that $a_j = \text{Re } a_j + \mathcal{O}(h)$, we see that

$$B_j(z) = A_j(z)^* A_j(z) + \psi_j^2 + hB_j'(z),$$

with $A_j(z) = U_j\varphi_j(A - z)U_j^{-1}\varphi_j$, and $B_j'(z)$ is a uniformly bounded h-admissible operator. As a consequence, if $v \in L^2(I\!R^n; \mathcal{H})$, and for $\text{Im } z \neq 0$, a computation similar to that of (4.7) shows that,

$$B_j(z)U_j\chi_j u_j(z) = C_j(z)v + hC_j'(z)U_j\varphi_j u_j(z) + \mathcal{O}(h^\infty)\|v\|, \tag{4.9}$$

where $u_j(z) := (A - z)^{-1}U_j^{-1}\chi_j v$, and $C_j(z), C_j'(z)$ are uniformly bounded h-admissible operators. Then, denoting by $\tilde{B}_j(z)$ the h-admissible operator with symbol $[(\text{ Re } a_j - \overline{z})(\text{ Re } a_j - z)\varphi_j^4 + \psi_j^2]^{-1}$, the standard pseudodifferential calculus with operator-valued symbols shows that,

$$\|\tilde{B}_j(z)\| = \mathcal{O}(|\text{ Im } z|^{-N_0})$$

for some $N_0 \geq 1$, and,

$$\tilde{B}_j(z)B_j(z) = 1 + hR_j(z),$$

where $R_j(z)$ is a h-admissible operator with symbol $r_j(z)$ verifying $\partial_{x,\xi}^\alpha r_j(z) = \mathcal{O}(|\text{ Im } z|^{-N_{\alpha,j}})$, for all $\alpha \in \mathbb{Z}_+^{2n}$, and for some $N_{\alpha,j} \geq 1$. Thus, applying $B_j'(z)$ to (4.9), we obtain,

$$U_j\chi_j u_j(z) = C_j^{(1)}(z)v + hC_j^{(2)}(z)U_j\varphi_j u_j(z) + \mathcal{O}(h^\infty|\text{ Im } z|^{-N_1})\|v\|,$$

where $C_j^{(1)}(z), C_j^{(2)}(z)$ are two h-admissible operators, uniformly bounded by some negative power of $|\text{ Im } z|$, and N_1 is some positive number. Again, iterating the procedure as in the proof of Proposition 4.10, one can deduce that $f(A)$ also verifies the second condition in Proposition 4.6, and therefore is a $\mathcal{U}$-twisted h-admissible operator.

Finally, a computation similar to that of (4.6) shows that,

$$(A - z)^{-1} = \sum_{j=0}^{r} U_j^{-1}\varphi_j^3\text{Op}_h(\varphi_j(\text{ Re } a_j - z)^{-1})U_j\chi_j + hR$$

where φ_j and χ_j are as in (4.6), and R verifies,

$$U_j\tilde{\chi}_j RU_j^{-1}\tilde{\chi}_j = \text{Op}_h(\sum_{k=0}^{N} h^k r_{k,j}(z)) + \mathcal{O}(h^N|\text{ Im } z|^{-N_1(N)}),$$

for any $\tilde{\chi}_j \in C_d^\infty(\Omega_j)$ such that $\tilde{\chi}_j\varphi_j = \tilde{\chi}_j\chi_j = \tilde{\chi}_j$, any $N \geq 1$, and for some $N_1(N) \geq 1$ and $r_{k,j}(z) \in C^\infty(T^*\Omega_j)$, $\partial^\alpha r_{k,j}(z) = \mathcal{O}(|\text{ Im } z|^{-N_{\alpha,k,j}})$ uniformly. Then, one easily concludes that the symbol b_j of $U_j\tilde{\chi}_j f(A)U_j\tilde{\chi}_j$ verifies,

$$b_j = \tilde{\chi}_j f(\text{ Re } a_j)\tilde{\chi}_j + \mathcal{O}(h),$$

and since the previous construction can be made for $\tilde{\chi}_j \in C_d^\infty(\Omega_j)$ arbitrary, the result on the symbol of $f(A)$ follows. •

In order to complete the theory of bounded $\mathcal{U}$-twisted h-admissible operators, it remains to generalize the notion of quantization. To this purpose, we first observe that, if $a = (a_j)_{j=0,\ldots,r} \in \mathbf{S}(\Omega; \mathcal{L}(H))$, then, the two operators $\varphi_j\text{Op}_h(a_j)\varphi_j$ and $U_j^{-1}\varphi_j\text{Op}_h(a_j)U_j\varphi_j$ are well defined for any $\varphi_j \in C_d^\infty(\Omega_j)$. Moreover, if $a = \sigma(A)$

is the symbol of a $\mathcal{U}$-twisted h-admissible operator A, then, by construction, it necessarily verifies the following condition of compatibility:

$$U_j^{-1}\varphi \mathrm{Op}_h(a_j)U_j\varphi = U_k^{-1}\varphi \mathrm{Op}_h(a_k)U_k\varphi, \tag{4.10}$$

for any $\varphi \in C_d^\infty(\Omega_j)\cap C_d^\infty(\Omega_k)$. Then, we have,

THEOREM 4.13 (Quantization). *Let $a = (a_j)_{j=0,\dots,r} \in \mathbf{S}(\Omega;\mathcal{L}(H))$ satisfying to the compatibility condition (4.10). Then, there exists a $\mathcal{U}$-twisted h-admissible operator A, unique up to $\mathcal{O}(h^\infty)$, such that $a=\sigma(A)$. Moreover, A is given by the formula,*

$$A = \sum_{j=0}^{r} U_j^{-1}\chi_j \mathrm{Op}_h(a_j)U_j\varphi_j, \tag{4.11}$$

where $\chi_j, \varphi_j \in C_d^\infty(\Omega_j) (j=0,\dots,r)$ is any family of functions such that $\sum_{j=0}^r \chi_j = 1$ and $\mathrm{dist}\,(\,\mathrm{Supp}\,(\varphi_j-1),\ \mathrm{Supp}\,\chi_j) > 0$.

Proof – The unicity up to $\mathcal{O}(h^\infty)$ is a direct consequence of the formulas (4.2)-(4.3), where A is expressed in terms of $U_j\varphi_j A U_j^{-1}\varphi_j$ and is clearly $\mathcal{O}(h^\infty)$ if these operators have identically vanishing symbols. For the existence, we define A as in (4.11) and we observe that, thanks to (4.10), for any $k\in\{0,\dots,r\}$ and $\psi_k \in C_d^\infty(\Omega_k)$, one has,

$$\begin{aligned} U_k\psi_k A U_k^{-1}\psi_k &= \sum_{j=0}^{r}\chi_j\psi_k \mathrm{Op}_h(a_k)\varphi_j\psi_k = \sum_{j=0}^{r}\chi_j\psi_k\mathrm{Op}_h(a_k)\psi_k + \mathcal{O}(h^\infty)\\ &= \psi_k\mathrm{Op}_h(a_k)\psi_k + \mathcal{O}(h^\infty). \end{aligned}$$

Thus, A admits $(a_k)_{k=0,\dots,r}$ as its symbol, and the result follows. •

To end this chapter, let us go back to our operator $\tilde{P}$ defined at the end of Chapter 3. We have,

PROPOSITION 4.14. *Assume (H1)-(H4). Then, the operator $\tilde{P}$ defined in (3.4) is such that $\tilde{P}(\omega+Q_0)^{-1}$ is a $\mathcal{U}$-twisted h-admissible operator on $L^2(\mathbb{R}^n;\mathcal{H})$, where $\mathcal{U} = (U_j,\Omega_j)_{j=0,1,\cdots,r}$ is the regular covering defined in Chapter 2. Moreover, its symbol $\tilde{p} = (\tilde{p}_j)_{j=0,1,\cdots,r}$ verifies,*

$$\tilde{p}_j(x,\xi) = (\omega(x,\xi)+\tilde{Q}_j(x)+\zeta(x)W(x))(\omega(x,\xi)+Q_{0,j}(x))^{-1} + hb_j,$$

where $(\tilde{Q}_j(x))_{j=0,1,\cdots,r}$ (resp. $(Q_{0,j}(x))_{j=0,1,\cdots,r}$) is the symbol of $\tilde{Q}(x)$ (resp. $Q_0(x)$), and $(b_j)_{j=0,\dots,r}\in \mathbf{S}(\Omega;\mathcal{L}(H))$.

Proof – We must verify the two conditions of Proposition 4.6. We have,

$$\begin{aligned} &\mathrm{ad}_\chi(\tilde{P}(\omega+Q_0)^{-1})\\ &= \mathrm{ad}_\chi(\tilde{P})(\omega+Q_0)^{-1} + \tilde{P}\mathrm{ad}_\chi((\omega+Q_0)^{-1})\\ &= \mathrm{ad}_\chi(\omega)(\omega+Q_0)^{-1} - \tilde{P}(\omega+Q_0)^{-1}\mathrm{ad}_\chi(\omega)(\omega+Q_0)^{-1}\\ &= \mathcal{O}(h), \end{aligned}$$

and an easy iteration shows that the first condition of Proposition 4.6 is satisfied. Moreover, if $\chi_j, \tilde{\chi}_j \in C_b^\infty(\mathbb{R}^n)$ are supported in Ω_j $(j=1\cdots,r)$ and verify

$\text{Supp}\,\chi_j \cap \text{Supp}\,(1-\tilde{\chi}_j) = \emptyset$, and if we set $P_j := U_j\chi_j\tilde{P}U_j^{-1}\tilde{\chi}_j$, we have,

$$\begin{aligned} &U_j\chi_j\tilde{P}(\omega+Q_0)^{-1}U_j^{-1}\chi_j \\ &\quad = U_j\chi_j\tilde{P}\tilde{\chi}_j^2(\omega+Q_0)^{-1}U_j^{-1}\chi_j + U_j\chi_j\omega(1-\tilde{\chi}_j^2)(\omega+Q_0)^{-1}U_j^{-1}\chi_j \\ &\quad = P_jU_j\tilde{\chi}_j(\omega+Q_0)^{-1}U_j^{-1}\chi_j + \mathcal{O}(h^\infty), \end{aligned}$$

and a slight generalization of the last argument in the proof of Proposition 4.10 (this time with $B_j = U_j\varphi_j(\omega+Q_0)U_j^{-1}\varphi_j + \psi_j(\omega+Q_0)\psi_j$), shows that $P_jU_j\tilde{\chi}_j(\omega+Q_0)^{-1}U_j^{-1}\chi_j$ is a bounded h-admissible operator on $L^2(\mathbb{R}^n;\mathcal{H})$. Therefore, the second condition of Proposition 4.6 is satisfied, too, and the result follows. •

COROLLARY 4.15. *The two operators $(\tilde{P}+i)^{-1}$ and $(\omega+Q_0)^{-1}$ are $\mathcal{U}$-twisted h-admissible operators on $L^2(\mathbb{R}^n;\mathcal{H})$.*

Proof – First observe that the previous proof is still valid if $\tilde{P}$ is changed into $\tilde{P}+1$. This proves that $(\omega+Q_0)^{-1} = (\tilde{P}+1)(\omega+Q_0)^{-1} - \tilde{P}(\omega+Q_0)^{-1}$ is a $\mathcal{U}$-twisted h-admissible operator. Moreover, since $(\tilde{P}+i)(\omega+Q_0)^{-1}$ is elliptic, by Proposition 4.10 its inverse $(\omega+Q_0)(\tilde{P}+i)^{-1}$ is a $\mathcal{U}$-twisted h-admissible operator, too. Therefore, so is $(\tilde{P}+i)^{-1} = (\omega+Q_0)^{-1}\left[(\omega+Q_0)(\tilde{P}+i)^{-1}\right]$. •

PROPOSITION 4.16. *For any $f \in C_0^\infty(\mathbb{R})$, the operator $f(\tilde{P})$ is a $\mathcal{U}$-twisted h-admissible.*

Proof – By Proposition 4.14 and Corollary 4.15, we see that the operator $(\tilde{P}-z)(\omega+Q_0)^{-1}$ is a $\mathcal{U}$-twisted h-admissible operator, and it is elliptic for $z \in \mathbb{C}\setminus\mathbb{R}$. Therefore, by Proposition 4.10, its inverse $(\omega+Q_0)(\tilde{P}-z)^{-1}$ is a $\mathcal{U}$-twisted h-admissible operator, too. Moreover, for any $N\geq 1$ and any $\chi_1,\cdots,\chi_N \in C_b^\infty(\mathbb{R}^n)$, one has,

$$\text{ad}_{\chi_1}\circ\cdots\circ\text{ad}_{\chi_N}((\omega+Q_0)(\tilde{P}-z)^{-1}) = \mathcal{O}(h^N|\,\text{Im}\,z|^{-(N+1)})$$

uniformly with respect to h and z. Therefore, we deduce again from (4.8) that $(\omega+Q_0)f(\tilde{P})$, too, is a $\mathcal{U}$-twisted h-admissible operator. As a consequence, so is $f(\tilde{P})$. •

CHAPTER 5

Twisted Partial Differential Operators

For $\mu \geq 0$, we set,

$$H_d^\mu(\Omega_j) := \{u \in L^2(\Omega_j;\mathcal{H})\,;\, \forall \chi_j \in C_d^\infty(\Omega_j),\ \chi_j u \in H^\mu(\mathbb{R}^n;\mathcal{H})\},$$

where $H^\mu(\mathbb{R}^n;\mathcal{H})$ stands for the usual Sobolev space of order μ on $\mathbb{R}^n$ with values in $\mathcal{H}$. Moreover, if $\mathcal{U} := (U_j,\Omega_j)_{j=0,\cdots,r}$ is a regular unitary covering (in the previous sense) of $L^2(\mathbb{R}^n;\mathcal{H})$, we introduce the vector-space,

$$\mathcal{H}_d^\mu(\mathcal{U}) := \{u \in L^2(\mathbb{R}^n;\mathcal{H})\,;\, \forall\, j = 0,\ldots,r,\ U_j u\left|_{\Omega_j}\right. \in H_d^\mu(\Omega_j)\},$$

endowed with the family of semi-norms,

$$\|u\|_{\mu,\chi} := \|u\|_{L^2} + \sum_{j=0}^{r} \|U_j\chi_j u\|_{H^\mu},$$

where $\chi := (\chi_j)_{j=0,\ldots,r}$ is such that $\chi_j \in C_d^\infty(\Omega_j)$ for all j. In particular, we have a notion of continuity for operators $A : \mathcal{H}_d^\mu(\mathcal{U}) \to \mathcal{H}_d^\nu(\mathcal{U})$.

Let us also remark that, for $\mu = 0$, we recover $\mathcal{H}_d^0(\mathcal{U}) = L^2(\mathbb{R}^n;\mathcal{H})$, and, if $\mu \geq \nu$, then $\mathcal{H}_d^\mu(\mathcal{U}) \subset \mathcal{H}_d^\nu(\mathcal{U})$ with a continuous injection.

DEFINITION 5.1. *Let $\mathcal{U} := (U_j,\Omega_j)_{j=0,\cdots,r}$ be a regular unitary covering (in the previous sense) of $L^2(\mathbb{R}^n;\mathcal{H})$, and let $\mu \in \mathbb{Z}_+$. We say that an operator $A : \mathcal{H}_d^\mu(\mathcal{U}) \to L^2(\mathbb{R}^n;\mathcal{H})$ is a (semiclassical) $\mathcal{U}$-twisted partial differential operator up to regularizing unitary conjugation (in short: $\mathcal{U}$-twisted PDO) of degree μ, if A is local with respect to the variable x (that is, $\mathrm{Supp}\,(Au) \subset \mathrm{Supp}\,u$ for all u, where Supp stands for the support with respect to x), and, for all $j = 0,\ldots,r$, the operator $U_jAU_j^{-1}$ (well defined on $H_d^\mu(\Omega_j)$) is of the form,*

$$U_jAU_j^{-1} = \sum_{|\alpha|\leq\mu} a_{\alpha,j}(x;h)(hD_x)^\alpha$$

with $a_{\alpha,j} \in S(\Omega_j;\mathcal{L}(\mathcal{H}))$.

In particular, for any partition of unity $(\chi_j)_{j=0,\ldots,r}$ on $\mathbb{R}^n$ with $\chi_j \in C_d^\infty(\Omega_j)$, A can be written as,

$$A = \sum_{j=0}^{r} U_j^{-1}A_jU_j\chi_j, \tag{5.1}$$

with $A_j := U_jAU_j^{-1}$. As a consequence, one also has $\mathrm{ad}_{\chi_1} \circ \cdots \circ \mathrm{ad}_{\chi_{\mu+1}}(A) = 0$ for any functions $\chi_1,\cdots,\chi_{\mu+1} \in C_b^\infty(\mathbb{R}^n)$.

Of course, we also have an obvious notion of (full) symbol for such operators, namely, the family,

$$\sigma(A) := (a_j)_{0\leq j\leq r}, \quad a_j(x,\xi;h) := \sum_{|\alpha|\leq\mu} a_{\alpha,j}(x;h)\xi^\alpha.$$

Moreover, if A and B are two $\mathcal{U}$-twisted PDO's on $L^2(\mathbb{R}^n;\mathcal{H})$, of respective degrees μ and μ', by writing $U_jABU_j^{-1} = (U_jAU_j^{-1})(U_jBU_j^{-1})$ and by using a partition of unity as before, we immediately see that AB is well defined on $\mathcal{H}_d^{\mu+\mu'}(\mathcal{U})$, and is a $\mathcal{U}$-twisted PDO, too, with symbol,

$$\sigma(AB) = \sigma(A)\sharp\sigma(B).$$

Now, we turn back again to the operator $\tilde{P}$ defined at the end of Chapter 3, and the regular covering defined in Chapter 2.

PROPOSITION 5.2. *Let A be a $\mathcal{U}$-twisted PDO on $L^2(\mathbb{R}^n;\mathcal{H})$ of degree μ, where $\mathcal{U}$ is the regular covering defined in Chapter 2. Then, for any integers k,ℓ such that $k+\ell\geq\mu/m$, the operator $(\tilde{P}+i)^{-k}A(\tilde{P}+i)^{-\ell}$ is a $\mathcal{U}$-twisted h-admissible operator.*

Proof – We first consider the case $k=0$. For $\varphi_j,\psi_j\in C_d^\infty(\Omega_j)$, such that dist (Supp (ψ_j-1), Supp φ_j) >0, we have,

$$U_j\varphi_jA(\tilde{P}+i)^{-\ell}U_j^{-1}\varphi_j = U_j\varphi_jAU_j^{-1}\psi_jU_j\psi_j(\tilde{P}+i)^{-\ell}U_j^{-1}\varphi_j, \tag{5.2}$$

and, as in the proof of Proposition 4.10, we see that the inverse of $(\tilde{P}+i)^\ell$ can be written as,

$$(\tilde{P}+i)^{-\ell} = B(1+hR) \tag{5.3}$$

where R is uniformly bounded, and B is of the form,

$$B = \sum_{\nu=0}^{r} U_\nu^{-1}\tilde{\chi}_\nu\mathrm{Op}_h((p_\nu+i)^{-\ell})U_\nu\chi_\nu, \tag{5.4}$$

where $(\chi_\nu)_{\nu=0,\dots,r}$ is an arbitrary partition of unity with $\chi_\nu\in C_d^\infty(\Omega_\nu)$, $\tilde{\chi}_\nu\in C_d^\infty(\Omega_\nu)$ is such that $\tilde{\chi}_\nu\chi_\nu=\chi_\nu$, and $p_\nu(x,\xi;h)=\omega(x,\xi;h)+\tilde{Q}_\nu(x)+\zeta(x)W(x)$.

LEMMA 5.3. *Let $j\in\{0,\dots,r\}$ and $\psi_j\in C_d^\infty(\Omega_j)$ be fixed. Then, there exists a partition of unity $(\chi_\nu)_{\nu=0,\dots,r}$ of $\mathbb{R}^n$ with $\chi_\nu\in C_d^\infty(\Omega_\nu)$, and there exists $\tilde{\chi}_\nu\in C_d^\infty(\Omega_\nu)$ with $\tilde{\chi}_\nu\chi_\nu=\chi_\nu$ $(\nu=0,\dots,r)$, such that $\chi_j\psi_j=\psi_j$ and $\tilde{\chi}_\nu\psi_j=0$ if $\nu\neq j$.*

Proof – It is enough to construct a partition of unity in such a way that dist (Supp ψ_j, Supp (χ_j-1)) >0 (and thus, automatically, one will also have dist (Supp ψ_j, Supp χ_ν) >0 for $\nu\neq j$). Let $(\chi'_\nu)_{\nu=0,\dots,r}$ be a partition of unity as in Definition 4.1, and let $\chi''_j\in C_d^\infty(\Omega_j;[0,1])$ such that $\chi''_j=1$ in a neighborhood of Supp $\psi_j\cup$ Supp χ_j. Then, the result is obtained by taking $\chi_\nu:=(1-\chi''_j)\chi'_\nu$ if $\nu\neq j$, and $\chi_j:=\chi''_j$. •

Taking the χ_ν's and $\tilde{\chi}_\nu$'s as in the previous lemma, we obtain from (5.3)-(5.4),

$$U_j\psi_j(\tilde{P}+i)^{-\ell} = \psi_j\mathrm{Op}_h((p_j+i)^{-\ell})U_j\chi_j(1+hR),$$

and thus, since $U_j\varphi_j AU_j^{-1}\psi_j$ is a differential operator of degree μ with operator-valued symbol, we easily deduce from (5.2) that if $m\ell \geq \mu$, then $A(\tilde{P}+i)^{-\ell}$ is bounded on $L^2(\mathbb{R}^n;\mathcal{H})$, uniformly with respect to $h>0$. Moreover, writing,

$$U_j\varphi_j A(\tilde{P}+i)^{-\ell}U_j^{-1}\varphi_j = [U_j\varphi_j AU_j^{-1}\psi_j\langle hD_x\rangle^{-m\ell}][\langle hD_x\rangle^{m\ell}U_j\psi_j(\tilde{P}+i)^{-\ell}U_j^{-1}\varphi_j],$$

and using the standard pseudodifferential calculus with operator-valued symbol for the first factor, and a slight refinement of (4.7) for the second one, we see that $U_j\varphi_j A(\tilde{P}+i)^{-\ell}U_j^{-1}\varphi_j$ is an h-admissible operator on $L^2(\mathbb{R}^n);\mathcal{H})$. Then, it only remains to verify the first property of Proposition 4.6. We first prove,

LEMMA 5.4. *For any $\alpha_1,\dots,\alpha_N \in C_b^\infty(\mathbb{R}^n)$, one has,*

$$\mathrm{ad}_{\alpha_1}\circ\cdots\circ\mathrm{ad}_{\alpha_N}((\tilde{P}+i)^{-\ell}) = h^N(\tilde{P}+i)^{-\ell}R_N, \tag{5.5}$$

with $R_N=\mathcal{O}(1)$ on $L^2(\mathbb{R}^n;\mathcal{H})$.

Proof – Since $\mathrm{ad}_{\alpha_N}((\tilde{P}+i)^{-\ell}) = -(\tilde{P}+i)^{-\ell}\mathrm{ad}_{\alpha_N}((\tilde{P}+i)^{\ell})(\tilde{P}+i)^{-\ell}$, by an easy iteration we see that it is enough to prove that $h^{-N}\mathrm{ad}_{\alpha_1}\circ\cdots\circ\mathrm{ad}_{\alpha_N}((\tilde{P}+i)^{\ell})(\tilde{P}+i)^{-\ell}$ is uniformly bounded. Moreover, since $\mathrm{ad}_{\alpha_N}((\tilde{P}+i)^{\ell})(\tilde{P}+i)^{-\ell}$ is a sum of terms of the type $(\tilde{P}+i)^k\mathrm{ad}_{\alpha_N}(\boldsymbol{\omega})(\tilde{P}+i)^{-k-1}$ $(0\leq k\leq \ell-1)$, another easy iteration shows that it is enough to prove that $h^{-N}(\tilde{P}+i)^{\ell}\mathrm{ad}_{\alpha_1}\circ\cdots\circ\mathrm{ad}_{\alpha_N}(\boldsymbol{\omega})(\tilde{P}+i)^{-\ell-1}$ is uniformly bounded. Now, by (H4), we see that, for any partition of unity (χ_j) as before, $(\tilde{P}+i)^{\ell}$ can be written as,

$$(\tilde{P}+i)^{\ell} = \sum_{j=0}^{r} U_j^{-1}P_{j,\ell}U_j\chi_j,$$

where $P_{j,\ell}$ is of the form,

$$P_{j,\ell} = \sum_{|\alpha|\leq m\ell} \rho_{j,\ell,\alpha}(x;h)(hD_x)^{\alpha},$$

with $\rho_{j,\ell,\alpha}Q_0^{\frac{|\alpha|}{m}-\ell} \in C^\infty(\Omega_j;\mathcal{H})$. Moreover, by (2.3), the operator $U_j\mathrm{ad}_{\alpha_1}\circ\cdots\circ\mathrm{ad}_{\alpha_N}(\boldsymbol{\omega})U_j^{-1} = \mathrm{ad}_{\alpha_1}\circ\cdots\circ\mathrm{ad}_{\alpha_N}(U_j\boldsymbol{\omega}U_j^{-1})$ is of the form,

$$U_j\mathrm{ad}_{\alpha_1}\circ\cdots\circ\mathrm{ad}_{\alpha_N}(\boldsymbol{\omega})U_j^{-1} = h^N\sum_{|\alpha|\leq(m-N)_+}\tau_{j,\alpha}(x;h)(hD_x)^{\alpha},$$

with $\tau_{j,\alpha}Q_0^{\frac{|\alpha|}{m}-1}\in C^\infty(\Omega_j;\mathcal{H})$. In particular, we obtain,

$$(\tilde{P}+i)^{\ell}\mathrm{ad}_{\alpha_1}\circ\cdots\circ\mathrm{ad}_{\alpha_N}(\boldsymbol{\omega}) = h^N\sum_{j=0}^{r}\sum_{|\alpha|\leq m(\ell+1)} U_j^{-1}\lambda_{j,\ell,\alpha}(x;h)(hD_x)^{\alpha}U_j\varphi_j,$$

with $\varphi_j\in C_d^\infty(\Omega_j)$ and $\lambda_{j,\ell,\alpha}Q_0^{\frac{|\alpha|}{m}-\ell-1}\in C^\infty(\Omega_j;\mathcal{H})$, and the result follows as before by using (5.3)-(5.4), and by observing that, for $|\alpha|\leq m(\ell+1)$, the operator $Q_0^{1+\ell-\frac{|\alpha|}{m}}(hD_x)^{\alpha}(\langle hD_x\rangle^m+Q_0)^{-\ell-1}$ is uniformly bounded, and thus so is the operator $Q_0^{1+\ell-\frac{|\alpha|}{m}}(hD_x)^{\alpha}\varphi_j\mathrm{Op}_h((p_j+i)^{-\ell-1})U_j\chi_j$. •

On the other hand, we see on (5.1) that $\mathrm{ad}_{\chi_1}\circ\cdots\circ\mathrm{ad}_{\chi_N}(A)$ is a $\mathcal{U}$-twisted PDO of degree $(\mu-N)_+$, and the first property of Proposition 4.6 for $A(\tilde{P}+i)^{-\ell}$ follows easily.

For the case $k > 0$, by taking a partition of unity, we first observe that,

$$(\tilde{P}+i)^{-k}A(\tilde{P}+i)^{-\ell} = \sum_{j=0}^{r}(\tilde{P}+i)^{-k}U_j^{-1}A_jU_j\chi_j(\tilde{P}+i)^{-\ell}$$

where $A_j = U_jAU_j^{-1}$ can be written as,

$$A_j = \sum_{\substack{|\alpha|\leq mk\\|\beta|\leq m\ell}} (hD_x)^{\alpha}a_{\alpha,\beta,j}(x;h)(hD_x)^{\beta}.$$

Then, by using (in addition to (5.3)-(5.4)) that,

$$(\tilde{P}+i)^{-k} = (1+hR')B'$$

where R' is uniformly bounded, and B' is of the form,

$$B' = \sum_{\nu=0}^{r} U_{\nu}^{-1}\chi_{\nu}\mathrm{Op}_h((p_{\nu}+i)^{-\ell})U_{\nu}\tilde{\chi}_{\nu},$$

the same previous arguments show that the operator $(\tilde{P}+i)^{-k}A(\tilde{P}+i)^{-\ell}$ is bounded on $L^2(\mathbb{R}^n;\mathcal{H})$, uniformly with respect to $h > 0$.

Then, let $N \geq 1$ and $\alpha_1 \dots, \alpha_N \in C_d^{\infty}(\Omega_j)$, such that $\alpha_1\varphi_j = \varphi_j$, $\alpha_2\alpha_1 = \alpha_1$, ..., $\alpha_N\alpha_{N-1} = \alpha_{N-1}$, and $\alpha_N(\psi_j - 1) = 0$. We have,

$$\begin{aligned}
&U_j\varphi_j(\tilde{P}+i)^{-k}A(\tilde{P}+i)^{-\ell}U_j^{-1}\varphi_j\\
&\quad= U_j\varphi_j(\tilde{P}+i)^{-k}A\psi_j(\tilde{P}+i)^{-\ell}U_j^{-1}\varphi_j\\
&\qquad+U_j\varphi_j(\tilde{P}+i)^{-k}A(\psi_j-1)\mathrm{ad}_{\alpha_1}\circ\cdots\circ\mathrm{ad}_{\alpha_N}((\tilde{P}+i)^{-\ell})U_j^{-1}\varphi_j
\end{aligned}$$

and thus, by (5.5),

$$\begin{aligned}
&U_j\varphi_j(\tilde{P}+i)^{-k}A(\tilde{P}+i)^{-\ell}U_j^{-1}\varphi_j\\
&\qquad= U_j\varphi_j(\tilde{P}+i)^{-k}A\psi_j(\tilde{P}+i)^{-\ell}U_j^{-1}\varphi_j + \mathcal{O}(h^N).
\end{aligned}$$

Then, writing $A\psi_j = U_j^{-1}\tilde{\psi}_jA_jU_j\psi_j$, with $A_j = U_jAU_j^{-1}$ and $\tilde{\psi}_j \in C_d^{\infty}(\Omega_j)$ such that $\tilde{\psi}_j\psi_j = \psi_j$, the result is obtained along the same lines as before. •

PROPOSITION 5.5. *The two operators ωQ_0^{-1} and $Q_0^{-1}\omega$ are $\mathcal{U}$-twisted PDO's of degree m. Moreover, if A is a $\mathcal{U}$-twisted PDO such that Q_0A and AQ_0 are $\mathcal{U}$-twisted PDO's, too, of degree μ, then the operator $h^{-1}[\omega, A]$ is a $\mathcal{U}$-twisted PDO of degree at most $\mu + m - 1$.*

Proof – Thank to (H4), the fact that ωQ_0^{-1} and $Q_0^{-1}\omega$ are $\mathcal{U}$-twisted PDO's of degree m is obvious. Moreover, the fact that Q_0A and AQ_0 are both $\mathcal{U}$-twisted PDO's implies that $U_jAU_j^{-1}$ can be written as,

$$U_jAU_j^{-1} = \sum_{|\alpha|\leq\mu} a_{\alpha,j}(x;h)(hD_x)^{\alpha}$$

with $Q_0 a_{\alpha,j}$ and $a_{\alpha,j} Q_0$ in $S(\Omega_j; \mathcal{L}(\mathcal{H}))$. Then, using (H4), we have,

$$\begin{aligned} U_j \boldsymbol{\omega} A U_j^{-1} &- \sum_{\substack{|\alpha|\leq m \\ |\beta|\leq \mu}} c_\alpha(x;h)(hD_x)^\alpha a_{\beta,j}(x;h)(hD_x)^\beta \\ &\qquad + h \sum_{\substack{|\alpha|\leq m-1 \\ |\beta|\leq \mu}} \omega_{\alpha,j}(x;h)(hD_x)^\alpha a_{\beta,j}(x;h)(hD_x)^\beta \end{aligned}$$

and

$$\begin{aligned} U_j A \boldsymbol{\omega} U_j^{-1} &= \sum_{\substack{|\alpha|\leq m \\ |\beta|\leq \mu}} a_{\beta,j}(x;h)(hD_x)^\beta c_\alpha(x;h)(hD_x)^\alpha \\ &\qquad + h \sum_{\substack{|\alpha|\leq m-1 \\ |\beta|\leq \mu}} a_{\beta,j}(x;h)(hD_x)^\beta \omega_{\alpha,j}(x;h)(hD_x)^\alpha. \end{aligned}$$

Moreover, by (H4) (and the fact that $U_j \boldsymbol{\omega} U_j^{-1}$ is symmetric), we know that c_α is scalar-valued, and $Q_0^{-1}\omega_{\alpha,j}$, $\omega_{\alpha,j} Q_0^{-1}$ are bounded operators on $\mathcal{H}$ (together with all their derivatives). Thus, it is clear that $h^{-1} U_j [\boldsymbol{\omega}, A] U_j^{-1}$ is a PDO of degree $\leq \mu + m - 1$, and the result follows. •

CHAPTER 6

Construction of a Quasi-Invariant Subspace

THEOREM 6.1. *Assume (H1)-(H4), and denote by* $\mathcal{U} := (U_j, \Omega_j)_{j=0,\cdots,r}$ *the regular unitary covering of* $L^2(I\!\!R^n;\mathcal{H})$ *constructed from the operators* U_j *and the open sets* Ω_j *defined in Chapter 2. Then, for any* $g \in C_0^\infty(I\!\!R)$*, there exists a* $\mathcal{U}$*-twisted* h*-admissible operator* Π_g *on* $L^2(I\!\!R^n;\mathcal{H})$*, such that* Π_g *is an orthogonal projection that verifies,*

$$\Pi_g = \tilde{\Pi}_0 + \mathcal{O}(h) \tag{6.1}$$

and, for any $f \in C_0^\infty(I\!\!R)$ *with* $\operatorname{Supp} f \subset \{g = 1\}$*, and any* $\ell \geq 0$*,*

$$\tilde{P}^\ell[f(\tilde{P}), \Pi_g] = \mathcal{O}(h^\infty). \tag{6.2}$$

Moreover, Π_g *is uniformly bounded as an operator :* $L^2(I\!\!R^n;\mathcal{H}) \to L^2(I\!\!R^n;\mathcal{D}_Q)$ *and, for any* $\ell \geq 0$*, any* $N \geq 1$*, and any functions* $\chi_1, \cdots, \chi_N \in C_b^\infty(I\!\!R^n)$*, one has,*

$$\tilde{P}^\ell \mathrm{ad}_{\chi_1} \circ \cdots \circ \mathrm{ad}_{\chi_N}(\Pi_g) = \mathcal{O}(h^N). \tag{6.3}$$

Proof – : We first perform a formal construction, by essentially following a procedure taken from [**Ne1**] (see also [**BrNo**] in the case $L = 1$). In the sequel, all the twisted PDO's that are involved are associated with the regular covering $\mathcal{U}$ constructed in Chapter 2, and we will omit to specify it all the time. We say that a twisted PDO is symmetric when it is formally selfadjoint with respect to the scalar product in $L^2(I\!\!R^n;\mathcal{H})$.

Since $\boldsymbol{Q} = \tilde{Q}(x) + \zeta(x)W(x)$ commutes with $\tilde{\Pi}_0$, we have,

$$[\tilde{P}, \tilde{\Pi}_0] = [\boldsymbol{\omega}, \tilde{\Pi}_0]. \tag{6.4}$$

Moreover, denoting by $\gamma(x)$ a complex oriented single loop surrounding the set $\{\tilde{\lambda}_{L'+1}(x), \ldots, \tilde{\lambda}_{L'+L}(x)\}$ and leaving the rest of the spectrum of $\tilde{Q}(x)$ in its exterior, we have,

$$\tilde{\Pi}_0(x) = \frac{1}{2i\pi}\int_{\gamma(x)} (z - \tilde{Q}(x))^{-1} dz, \tag{6.5}$$

and thus, it results from Proposition 3.2 and assumption (H4) that $Q_0\tilde{\Pi}_0(x)$ is a $\mathcal{U}$-twisted PDO of degree 0. Therefore, applying Proposition 5.5, we immediately obtain,

$$[\tilde{P}, \tilde{\Pi}_0] = -ihS_0, \tag{6.6}$$

where S_0 is a symmetric twisted PDO (of degree $m-1$). Moreover, setting $\tilde{\Pi}_0^\perp := 1 - \tilde{\Pi}_0$, we observe that,

$$S_0 = \tilde{\Pi}_0 S_0 \tilde{\Pi}_0^\perp + \tilde{\Pi}_0^\perp S_0 \tilde{\Pi}_0. \tag{6.7}$$

Then, we set,
(6.8)

$$\tilde{\Pi}_1 := -\frac{1}{2\pi}\oint_{\gamma(x)} (z-\tilde{Q}(x))^{-1}\left[\tilde{\Pi}_0^\perp(x)S_0\tilde{\Pi}_0(x) - \tilde{\Pi}_0(x)S_0\tilde{\Pi}_0^\perp(x)\right](z-\tilde{Q}(x))^{-1}dz.$$

Thus, $\tilde{\Pi}_1$ is a symmetric $\mathcal{U}$-twisted PDO (of degree $m-1$), and is such that $Q_0\tilde{\Pi}_1$ is a twisted PDO, too. Therefore, using Proposition 5.5 again, we have,

$$[\tilde{P},\tilde{\Pi}_1] = [\boldsymbol{Q},\tilde{\Pi}_1] + hB,$$

where B is a twisted PDO (of degree $2(m-1)$). Then, using that $\tilde{Q}(x)(z-\tilde{Q}(x))^{-1} = (z-\tilde{Q}(x))^{-1}\tilde{Q}(x) = z(z-\tilde{Q}(x))^{-1} - 1$, one computes,

$$\begin{aligned}
[\tilde{Q}(x),\tilde{\Pi}_1] &= \frac{1}{2\pi}\oint_{\gamma(x)}\left[\tilde{\Pi}_0^\perp(x)S_0\tilde{\Pi}_0(x) - \tilde{\Pi}_0(x)S_0\tilde{\Pi}_0^\perp(x)\right](z-\tilde{Q}(x))^{-1}dz\\
&\quad -\frac{1}{2\pi}\oint_{\gamma(x)}(z-\tilde{Q}(x))^{-1}\left[\tilde{\Pi}_0^\perp(x)S_0\tilde{\Pi}_0(x) - \tilde{\Pi}_0(x)S_0\tilde{\Pi}_0^\perp(x)\right]dz\\
&= i\left[\tilde{\Pi}_0^\perp(x)S_0\tilde{\Pi}_0(x) - \tilde{\Pi}_0(x)S_0\tilde{\Pi}_0^\perp(x)\right]\tilde{\Pi}_0(x)\\
&\qquad -i\tilde{\Pi}_0(x)\left[\tilde{\Pi}_0^\perp(x)S_0\tilde{\Pi}_0(x) - \tilde{\Pi}_0(x)S_0\tilde{\Pi}_0^\perp(x)\right]\\
&= i(\tilde{\Pi}_0^\perp S_0\tilde{\Pi}_0 + \tilde{\Pi}_0 S_0\tilde{\Pi}_0^\perp),
\end{aligned}$$

that gives,

$$[\boldsymbol{Q},\tilde{\Pi}_1] = i(\tilde{\Pi}_0^\perp S_0\tilde{\Pi}_0 + \tilde{\Pi}_0 S_0\tilde{\Pi}_0^\perp) + [\zeta W,\tilde{\Pi}_1], \tag{6.9}$$

and thus, using (6.7), one obtains,

$$[\tilde{P},\tilde{\Pi}_1] = iS_0 - ihS_1, \tag{6.10}$$

where S_1 is a symmetric twisted PDO (of degree $2(m-1)$). Hence, setting,

$$\Pi^{(1)} := \tilde{\Pi}_0 + h\tilde{\Pi}_1,$$

we deduce from (6.6) and (6.10),

$$[\tilde{P},\Pi^{(1)}] = -ih^2S_1. \tag{6.11}$$

Moreover,

$$(\Pi^{(1)})^2 - \Pi^{(1)} = h(\tilde{\Pi}_0\tilde{\Pi}_1 + \tilde{\Pi}_1\tilde{\Pi}_0 - \tilde{\Pi}_1) + h^2\tilde{\Pi}_1^2 = h^2\tilde{\Pi}_1^2 =: h^2T_1,$$

where T_1 is a symmetric twisted PDO (of degree $2(m-1)$), such that Q_0T_1 is a twisted PDO, too.

Now, by induction on M, suppose that we have constructed a symmetric twisted PDO $\Pi^{(M)}$ as,

$$\Pi^{(M)} = \sum_{k=0}^{M} h^k\tilde{\Pi}_k,$$

where the $Q_0\tilde{\Pi}_k$'s are twisted PDO's, such that,

$$(\Pi^{(M)})^2 - \Pi^{(M)} = h^{M+1}T_M; \tag{6.12}$$

$$[\tilde{P},\Pi^{(M)}] = -ih^{M+1}S_M, \tag{6.13}$$

with S_M and Q_0T_M twisted PDO's.

We set,

$$\Pi^{(M+1)} = \Pi^{(M)} + h^{M+1}\tilde{\Pi}_{M+1},$$

with,

$$\begin{aligned}\tilde{\Pi}_{M+1} \quad &:= \quad -\frac{1}{2\pi}\oint_{\gamma(x)} (z-\tilde{Q}(x))^{-1}\left[\tilde{\Pi}_0^{\perp} S_M \tilde{\Pi}_0 - \tilde{\Pi}_0 S_M \tilde{\Pi}_0^{\perp}\right](z-\tilde{Q}(x))^{-1}dz \\ &\qquad + \tilde{\Pi}_0^{\perp} T_M \tilde{\Pi}_0^{\perp} - \tilde{\Pi}_0 T_M \tilde{\Pi}_0. \qquad (6.14)\end{aligned}$$

Then, $\Pi^{(M+1)}$ is again a symmetric twisted PDO, and, using the induction assumption, we immediately see that $\tilde{Q}(x)\tilde{\Pi}_{M+1}$ (and thus also $Q_0\tilde{\Pi}_{M+1}$) is a twisted PDO. Moreover, since T_M and $\Pi^{(M)}$ commute, we have,

$$\Pi^{(M)}T_M(1-\Pi^{(M)}) = (1-\Pi^{(M)})T_M\Pi^{(M)} = -h^{M+1}T_M^2,$$

and thus, since $\Pi^{(M)} = \tilde{\Pi}_0 + hR_M$ with Q_0R_M twisted PDO, we first obtain,

$$\tilde{\Pi}_0^{\perp} T_M \tilde{\Pi}_0 + \tilde{\Pi}_0 T_M \tilde{\Pi}_0^{\perp} = hR'_M, \tag{6.15}$$

with $Q_0R'_M$ twisted PDO. On the other hand, one can check that,

$$\tilde{\Pi}_{M+1} - (\tilde{\Pi}_0\tilde{\Pi}_{M+1} + \tilde{\Pi}_{M+1}\tilde{\Pi}_0) = \tilde{\Pi}_0 T_M \tilde{\Pi}_0 + \tilde{\Pi}_0^{\perp} T_M \tilde{\Pi}_0^{\perp},$$

and thus, with (6.15),

$$\tilde{\Pi}_{M+1} - (\tilde{\Pi}_0\tilde{\Pi}_{M+1} + \tilde{\Pi}_{M+1}\tilde{\Pi}_0) = T_M - hR'_M.$$

As a consequence, we obtain,

$$(\Pi^{(M+1)})^2 - \Pi^{(M+1)} = h^{M+2}T_{M+1}, \tag{6.16}$$

where Q_0T_{M+1} is a twisted PDO. Applying Proposition 5.5, we also have,

$$[\boldsymbol{\omega}, \tilde{\Pi}_{M+1}] = hR''_M,$$

with R''_M twisted PDO, and thus,

$$\begin{aligned}[\tilde{P}, \tilde{\Pi}_{M+1}] \quad &= \quad [\boldsymbol{Q}, \tilde{\Pi}_{M+1}] + hR''_M \\ &= \quad i(\tilde{\Pi}_0 S_M \tilde{\Pi}_0^{\perp} + \tilde{\Pi}_0^{\perp} S_M \tilde{\Pi}_0) \\ &\qquad + \tilde{\Pi}_0^{\perp}[\boldsymbol{Q}, T_M]\tilde{\Pi}_0^{\perp} - \tilde{\Pi}_0[\boldsymbol{Q}, T_M]\tilde{\Pi}_0 + hR_M^{(3)} \qquad (6.17)\end{aligned}$$

with $R_M^{(3)}$ twisted PDO, and, using the hypothesis of induction (and, again, the twisted symbolic calculus),

$$\begin{aligned}&\tilde{\Pi}_0^{\perp}[\boldsymbol{Q}, T_M]\tilde{\Pi}_0^{\perp} \\ &\quad = \tilde{\Pi}_0^{\perp}[\tilde{P}, T_M]\tilde{\Pi}_0^{\perp} + hR_M^{(4)} \\ &\quad = h^{-(M+1)}\tilde{\Pi}_0^{\perp}[\tilde{P}, (\Pi^{(M)})^2 - \Pi^{(M)}]\tilde{\Pi}_0^{\perp} + hR_M^{(4)} \\ &\quad = h^{-(M+1)}\tilde{\Pi}_0^{\perp}([\tilde{P}, \Pi^{(M)}]\Pi^{(M)} + \Pi^{(M)}[\tilde{P}, \Pi^{(M)}] - [\tilde{P}, \Pi^{(M)}])\tilde{\Pi}_0^{\perp} + hR_M^{(4)} \\ &\quad = -i\tilde{\Pi}_0^{\perp}(S_M\Pi^{(M)} + \Pi^{(M)}S_M - S_M)\tilde{\Pi}_0^{\perp} + hR_M^{(4)} \\ &\quad = i\tilde{\Pi}_0^{\perp} S_M \tilde{\Pi}_0^{\perp} + hR_M^{(5)}, \qquad (6.18)\end{aligned}$$

and, in the same way,

$$\tilde{\Pi}_0[\boldsymbol{Q}, T_M]\tilde{\Pi}_0 = -i\tilde{\Pi}_0 S_M \tilde{\Pi}_0 + hR_M^{(6)}, \tag{6.19}$$

where the operators $R_M^{(k)}$'s are all twisted PDO's. Inserting (6.18)-(6.19) into (6.17), we finally obtain,

$$[\tilde{P}, \tilde{\Pi}_{M+1}] = iS_M + hR_M^{(7)},$$

that implies,

$$[\tilde{P}, \Pi^{(M+1)}] = -ih^{M+2}S_{M+1},$$

where S_{M+1} is a twisted PDO. Therefore, the induction is established.

From this point, we follow an idea of [**So**]. Let $g \in C_0^\infty(\mathbb{R})$. Using Propositions 5.2 and 4.16, and writing $g(\tilde{P})\tilde{\Pi}_k = g(\tilde{P})(\tilde{P}+i)^N(\tilde{P}+i)^{-N}\tilde{\Pi}_k$, we see that the operators $g(\tilde{P})\tilde{\Pi}_k$ $(k \geq 0)$ are all twisted h-admissible operators. In particular, they are all bounded, uniformly with respect to h. Moreover, for any $\ell, \ell' \geq 0$, any $N \geq 1$, and any functions $\chi_1, \cdots, \chi_N \in C_b^\infty(\mathbb{R}^n)$, by construction, $h^{-N}\mathrm{ad}_{\chi_1} \circ \cdots \circ \mathrm{ad}_{\chi_N}(\tilde{\Pi}_k)$ is a twisted PDO, and thus, by Propositions 5.2 and 4.16, $h^{-N}\tilde{P}^\ell g(\tilde{P})\mathrm{ad}_{\chi_1} \circ \cdots \circ \mathrm{ad}_{\chi_N}(\tilde{\Pi}_k)\tilde{P}^{\ell'}$ is uniformly bounded. It is also easy to show (e.g., by using (6.24) hereafter) that,

$$\tilde{P}^\ell \mathrm{ad}_{\chi_1} \circ \cdots \circ \mathrm{ad}_{\chi_N}(g(\tilde{P}))\tilde{P}^{\ell'} = \mathcal{O}(h^N), \tag{6.20}$$

and therefore, we obtain,

$$h^{-N}\tilde{P}^\ell \mathrm{ad}_{\chi_1} \circ \cdots \circ \mathrm{ad}_{\chi_N}(g(\tilde{P})\tilde{\Pi}_k)\tilde{P}^{\ell'} = \mathcal{O}(1),$$

uniformly with respect to h. As a consequence, we can resum in a standard way the formal series of operators $\sum_{k=0}^\infty h^k g(\tilde{P})\tilde{\Pi}_k$ (see, e.g., [**Ma2**] Lemma 2.3.3), in such a way that, if we denote by $\Pi(g)$ such a resummation, we have,

$$\|\tilde{P}^\ell \mathrm{ad}_{\chi_1} \circ \cdots \circ \mathrm{ad}_{\chi_N}(\Pi(g) - \sum_{k=0}^{M-1} h^k g(\tilde{P})\tilde{\Pi}_k)\tilde{P}^{\ell'}\|_{\mathcal{L}(L^2(\mathbb{R}^n;\mathcal{H}))} = \mathcal{O}(h^{M+N}), \tag{6.21}$$

for any $\ell, \ell' \geq 0$, $M, N \geq 0$ and any $\chi_1, \cdots, \chi_N \in C_b^\infty(\mathbb{R}^n)$ (with the conventions $\mathrm{ad}_{\chi_1} \circ \cdots \circ \mathrm{ad}_{\chi_N}(\Pi(g)) = \Pi(g)$ if $N = 0$, and $\sum_{k=0}^{M-1} = 0$ if $M = 0$).

Then, we prove,

Lemma 6.2. *For any $\ell \geq 0$, one has,*

$$\|\tilde{P}^\ell(\Pi(g) - \Pi(g)^*)\|_{\mathcal{L}(L^2(\mathbb{R}^n;\mathcal{H}))} = \mathcal{O}(h^\infty). \tag{6.22}$$

Proof – In view of (6.21), it is enough to show that, for any $M \geq 1$, one has,

$$(\tilde{P}+i)^\ell [g(\tilde{P}), \Pi^{(M)}] = \mathcal{O}(h^{M+1}). \tag{6.23}$$

For $N \geq 1$ large enough, we set $g_N(s) := g(s)(s+i)^N \in C_0^\infty(\mathbb{R})$, and we observe that,

$$g(\tilde{P}) = g_N(\tilde{P})(\tilde{P}+i)^{-N} = \frac{1}{\pi}\int \bar{\partial}\tilde{g}_N(z)(\tilde{P}-z)^{-1}(\tilde{P}+i)^{-N} dz\, d\bar{z}, \tag{6.24}$$

where $\tilde{g}_N$ is an almost analytic extension of g_N. Therefore, we obtain,

$$\begin{aligned}&(\tilde{P}+i)^\ell[g(\tilde{P}), \Pi^{(M)}]\\ &= \frac{1}{\pi}\int \bar{\partial}\tilde{g}_N(z)(\tilde{P}-z)^{-1}(\tilde{P}+i)^{\ell-N}[\Pi^{(M)}, (\tilde{P}-z)(\tilde{P}+i)^N](\tilde{P}-z)^{-1}(\tilde{P}+i)^{-N} dz\, d\bar{z},\end{aligned} \tag{6.25}$$

and it follows from (6.13) and the twisted PDO calculus, that,

$$[\Pi^{(M)}, (\tilde{P}-z)(\tilde{P}+i)^N] = h^{M+1}R_{M,N} \tag{6.26}$$

where $R_{M,N}$ is a twisted PDO of degree $\mu_M + mN$, with μ_M the degree of S_M. Therefore, if we choose N such that $2mN - m\ell \geq \mu_M + mN$, that is, $N \geq \ell + \mu_M/m$, then (6.25)-(6.26) and Proposition 5.2 tell us that $h^{-(M+1)}[g(\tilde{P}), \Pi^{(M)}]$ is a twisted h-admissible operator, and the result follows. •

We set,

$$\tilde{\Pi}_g := \Pi(g) + \Pi(g)^* - \frac{1}{2}(g(\tilde{P}))\Pi(g)^* + \Pi(g)g(\tilde{P})) + (1-g(\tilde{P}))\tilde{\Pi}_0(1-g(\tilde{P})). \tag{6.27}$$

Then, $\tilde{\Pi}_g$ is a selfadjoint twisted h-admissible operator, and since $\Pi(g) = g(\tilde{P})\tilde{\Pi}_0 + \mathcal{O}(h)$, we have,

$$\|\tilde{\Pi}_g - \tilde{\Pi}_0\|_{\mathcal{L}(L^2(\mathbb{R}^n;\mathcal{H}))} + \|\tilde{\Pi}_g^2 - \tilde{\Pi}_g\|_{\mathcal{L}(L^2(\mathbb{R}^n;\mathcal{H}))} = \mathcal{O}(h). \tag{6.28}$$

By construction, we also have $\tilde{P}^\ell(g(\tilde{P})\Pi(g)^* - \Pi(g)g(\tilde{P})) = \mathcal{O}(h^\infty)$ for all $\ell \geq 0$, and thus, by Lemma 6.2,

$$\tilde{P}^\ell\tilde{\Pi}_g = \tilde{P}^\ell\left[\Pi(g) + (1-g(\tilde{P}))\left(\Pi(g) + \tilde{\Pi}_0(1-g(\tilde{P}))\right)\right] + \mathcal{O}(h^\infty). \tag{6.29}$$

Moreover, if $f \in C_0^\infty(\mathbb{R})$ is such that $\operatorname{Supp} f \subset \{g = 1\}$, and if we denote by $\Pi(f)$ a resummation of the formal series $\sum_{k\geq 0} h^k f(\tilde{P})\tilde{\Pi}_k$ as before, since $f(\tilde{P})(1-g(\tilde{P})) = 0$, $f(\tilde{P})\Pi(g) - \Pi(f) = \mathcal{O}(h^\infty)$, and $\tilde{P}^\ell(1-g(\tilde{P})\Pi(g)f(\tilde{P}) = \tilde{P}^\ell(1-g(\tilde{P})\Pi(g)^*f(\tilde{P}) + \mathcal{O}(h^\infty) = \tilde{P}^\ell(1-g(\tilde{P})\Pi(f) + \mathcal{O}(h^\infty) = \mathcal{O}(h^\infty)$, we deduce from (6.29) and Lemma 6.2,

$$\tilde{P}^\ell[f(\tilde{P}), \tilde{\Pi}_g] = \tilde{P}^\ell\left(\Pi(f) - \Pi(g)^*f(\tilde{P})\right) + \mathcal{O}(h^\infty) = \tilde{P}^\ell\left(\Pi(f) - \Pi(f)^*\right) + \mathcal{O}(h^\infty),$$

and thus,

$$\|\tilde{P}^\ell[f(\tilde{P}), \tilde{\Pi}_g]\|_{\mathcal{L}(L^2(\mathbb{R}^n;\mathcal{H}))} = \mathcal{O}(h^\infty). \tag{6.30}$$

On the other hand, we deduce from Lemma 6.2 and (6.12),

$$\begin{aligned} \tilde{P}^\ell(\Pi(g)^2 - \Pi(g^2)) &= \tilde{P}^\ell(\Pi(g)\Pi(g)^* - \Pi(g^2)) + \mathcal{O}(h^\infty) \\ &= \tilde{P}^\ell(\Pi(g)g(\tilde{P}) - \Pi(g^2)) + \mathcal{O}(h^\infty) \\ &= \mathcal{O}(h^\infty), \end{aligned} \tag{6.31}$$

and thus, using (6.29)-(6.31),

$$\tilde{P}^\ell(\tilde{\Pi}_g^2 - \tilde{\Pi}_g)f(\tilde{P}) = \mathcal{O}(h^\infty). \tag{6.32}$$

Then, following the arguments of [**Ne1, Ne2, NeSo, So**], for h small enough we can define the following orthogonal projection:

$$\Pi_g := \frac{1}{2i\pi}\int_{|z-1|=\frac{1}{2}} (\tilde{\Pi}_g - z)^{-1}\, dz, \tag{6.33}$$

and it verifies (see [**So**], Formula (3.9), and [**Ne1**], Proposition 3),

$$\Pi_g - \tilde{\Pi}_g = \frac{1}{2i\pi}(\tilde{\Pi}_g^2 - \tilde{\Pi}_g)\int_{|z-1|=\frac{1}{2}} (\tilde{\Pi}_g - z)^{-1}(2\tilde{\Pi}_g - 1)(1-\tilde{\Pi}_g - z)^{-1}(1-z)^{-1}\, dz. \tag{6.34}$$

In particular, we obtain from (6.32) and (6.34),

$$\tilde{P}^\ell(\Pi_g - \tilde{\Pi}_g)f(\tilde{P}) = \mathcal{O}(h^\infty), \tag{6.35}$$

and thus, we deduce from (6.28) and (6.30) that (6.1) and (6.2) hold.

In order to prove (6.3), we first observe that, by using (6.20), (6.21) and the fact that $\mathrm{ad}_{\chi_k}(\tilde\Pi_0) = 0$, we obtain,

$$\tilde P^\ell \mathrm{ad}_{\chi_1} \circ \cdots \circ \mathrm{ad}_{\chi_N}(\tilde\Pi_g) = \mathcal{O}(h^N), \tag{6.36}$$

for any $N \geq 1$. On the other hand, we have,

LEMMA 6.3. *For any $\ell \geq 0$ and $z \in \mathbb{C}$ such that $|z-1| = 1/2$, the operator $\tilde P^\ell(\tilde\Pi_g - z)^{-1}(\tilde P + i)^{-\ell}$ is uniformly bounded on $L^2(\mathbb{R}^n; \mathcal{H})$.*

Proof – Writing, for $\ell > 0$,

$$\begin{aligned} H_\ell : &= (\tilde P + i)^\ell (\tilde\Pi_g - z)^{-1}(\tilde P + i)^{-\ell} \\ &= H_{\ell-1} + (\tilde P + i)^{\ell-1}[\tilde P, (\tilde\Pi_g - z)^{-1}](\tilde P + i)^{-\ell} \\ &= H_{\ell-1} + H_{\ell-1}(\tilde P + i)^{\ell-1}[\tilde\Pi_g, \tilde P](\tilde P + i)^{-\ell} H_\ell, \end{aligned}$$

and performing an easy induction, we see that it is enough to prove that $(\tilde P + i)^{\ell-1}[\tilde\Pi_g, \tilde P](\tilde P + i)^{-\ell}$ is $\mathcal{O}(h)$. Due to (6.29), it is enough to study the two terms $(\tilde P + i)^{\ell-1}[\tilde\Pi(g), \tilde P](\tilde P + i)^{-\ell}$ and $(\tilde P + i)^{\ell-1}[\tilde\Pi_0, \tilde P](\tilde P + i)^{-\ell}$. By (6.13), the first one is $\mathcal{O}(h^\infty)$, while the second one is equal to $(\tilde P + i)^{\ell-1}[\tilde\Pi_0, \omega](\tilde P + i)^{-\ell}$ and thus, by Propositions 5.5 and 5.2, is $\mathcal{O}(h)$. •

Combining (6.36), (6.33) and Lemma 6.3, we easily obtain (6.3), and this completes the proof of Theorem 6.1. •

REMARK 6.4. *Observe that the previous proof also provides a way of computing the full symbol of $\tilde\Pi_g$ (and thus of Π_g, too) up to $\mathcal{O}(h^M)$, for any $M \geq 1$. Indeed, formulas (6.12), (6.13), and (6.14) permit to do it inductively.*

REMARK 6.5. *For this proof, we did not succeed in adapting the elegant argument of* [**Sj2**] *(as this was done for smooth interactions in* [**So**]*), because of a technical problem. Namely, this argument involves a translation in the spectral variable z, of the type $z \mapsto z + \omega(x,\xi)$, inside the symbol of the resolvent of $\tilde P$. In our case, this would have led to consider a symbol $\tilde a = (\tilde a_j)_{0 \leq j \leq r}$ of the type $\tilde a_j = a_j(x,\xi, z + \omega_j(x,\xi))$, where ω_j is the symbol of $U_j \omega U_j^{-1}$ and $a(x,\xi,z) = (a_j(x,\xi,z))_{0 \leq j \leq r}$ is the symbol of $(z - \tilde P)^{-1}$. But then, it is not clear to us (and probably may be wrong) that the compatibility conditions (4.10) are verified by $\tilde a$, and this prevents us from quantizing it in order to continue the argument.*

CHAPTER 7

Decomposition of the Evolution for the Modified Operator

In this chapter we prove a general result on the quantum evolution of $\tilde{P}$.

THEOREM 7.1. *Under the same assumtions as for Theorem 6.1, let $g \in C_0^\infty(I\!R)$. Then, one has the following results:*
1) Let $\varphi_0 \in L^2(I\!R^n;\mathcal{H})$ verifying,

$$\varphi_0 = f(\tilde{P})\varphi_0, \tag{7.1}$$

for some $f \in C_0^\infty(I\!R)$ such that $\operatorname{Supp} f \subset \{g=1\}$. Then, with the projection Π_g constructed in Theorem 6.1, one has,

$$e^{-it\tilde{P}/h}\varphi_0 = e^{-it\tilde{P}^{(1)}/h}\Pi_g\varphi_0 + e^{-it\tilde{P}^{(2)}/h}(1-\Pi_g)\varphi_0 + \mathcal{O}(|t|h^\infty\|\varphi_0\|) \tag{7.2}$$

uniformly with respect to h small enough, $t \in I\!R$ and φ_0 verifying (7.1), with,

$$\tilde{P}^{(1)} := \Pi_g\tilde{P}\Pi_g \quad ; \quad \tilde{P}^{(2)} := (1-\Pi_g)\tilde{P}(1-\Pi_g).$$

2) Let $\varphi_0 \in L^2(I\!R^n;\mathcal{H})$ (possibly h-dependent) verifying $\|\varphi_0\| = 1$, and,

$$\varphi_0 = f(\tilde{P})\varphi_0 + \mathcal{O}(h^\infty), \tag{7.3}$$

for some $f \in C_0^\infty(I\!R)$ such that $\operatorname{Supp} f \subset \{g=1\}$. Then, one has,

$$e^{-it\tilde{P}/h}\varphi_0 = e^{-it\tilde{P}^{(1)}/h}\Pi_g\varphi_0 + e^{-it\tilde{P}^{(2)}/h}(1-\Pi_g)\varphi_0 + \mathcal{O}(\langle t\rangle h^\infty) \tag{7.4}$$

uniformly with respect to h small enough and $t \in I\!R$.
3) There exists a bounded operator $\mathcal{W} : L^2(I\!R^n;\mathcal{H}) \to L^2(I\!R^n)^{\oplus L}$ with the following properties:

- *For any $j \in \{0,1,\dots,r\}$, and any $\varphi_j \in C_d^\infty(\Omega_j)$, the operator $\mathcal{W}_j := \mathcal{W}U_j^{-1}\varphi_j$ is an h-admissible operator from $L^2(I\!R^n;\mathcal{H})$ to $L^2(I\!R^n)^{\oplus L}$;*
- *$\mathcal{W}\mathcal{W}^* = 1$ and $\mathcal{W}^*\mathcal{W} = \Pi_g$;*
- *The operator $A := \mathcal{W}\tilde{P}\mathcal{W}^* = \mathcal{W}\tilde{P}^{(1)}\mathcal{W}^*$ is an h-admissible operator on $L^2(I\!R^n)^{\oplus L}$ with domain $H^m(I\!R^n)^{\oplus L}$, and its symbol $a(x,\xi;h)$ verifies,*

  $$a(x,\xi;h) = \omega(x,\xi;h)\mathbf{I}_L + \mathcal{M}(x) + \zeta(x)W(x)\mathbf{I}_L + hr(x,\xi;h)$$

  where $\mathcal{M}(x)$ is a $L\times L$ matrix depending smoothly on x, with spectrum $\{\tilde{\lambda}_{L'+1}(x),\dots,\tilde{\lambda}_{L'+L}(x)\}$, and $r(x,\xi:h)$ verifies,

  $$\partial^\alpha r(x,\xi;h) = \mathcal{O}(\langle\xi\rangle^{m-1})$$

  for any multi-index α and uniformly with respect to $(x,\xi) \in T^*I\!R^n$ and $h>0$ small enough.

In particular, $\mathcal{W}\big|_{\operatorname{Ran}\Pi_g} : \operatorname{Ran}\Pi_g \to L^2(I\!R^n)^{\oplus L}$ is unitary, and $e^{-it\tilde{P}^{(1)}/h}\Pi_g = \mathcal{W}^*e^{-itA/h}\mathcal{W}\Pi_g = \mathcal{W}^*e^{-itA/h}\mathcal{W}$ for all $t \in I\!R$.

REMARK 7.2. *In Chapter 10, we give a way of computing easily the expansion of A up to any power of h. As an example, we compute explicitly its first three terms (that is, up to $\mathcal{O}(h^4)$).*

Proof – 1) Setting $\varphi := e^{-it\tilde{P}/h}\varphi_0$, we have $f(\tilde{P})\varphi = \varphi$, and thus

$$ih\partial_t\Pi_g\varphi = \Pi_g\tilde{P}f(\tilde{P})\varphi = \Pi_g^2\tilde{P}f(\tilde{P})\varphi. \tag{7.5}$$

Moreover, writing $[\Pi_g, \tilde{P}]f(\tilde{P}) = [\Pi_g, \tilde{P}f(\tilde{P})] + \tilde{P}[f(\tilde{P}), \Pi_g]$, Theorem 6.1 tells us that $\|[\Pi_g, \tilde{P}]f(\tilde{P})\| = \mathcal{O}(h^\infty)$. Therefore, we obtain from (7.5),

$$ih\partial_t\Pi_g\varphi = \Pi_g\tilde{P}\Pi_g f(\tilde{P})\varphi + \mathcal{O}(h^\infty\|\varphi\|) = \tilde{P}^{(1)}\Pi_g\varphi + \mathcal{O}(h^\infty\|\varphi_0\|),$$

uniformly with respect to h and t. This equation can be re-written as,

$$ih\partial_t(e^{it\tilde{P}^{(1)}/h}\Pi_g\varphi) = \mathcal{O}(h^\infty\|\varphi_0\|),$$

and thus, integrating from 0 to t, we obtain,

$$\Pi_g\varphi = e^{-it\tilde{P}^{(1)}/h}\Pi_g\varphi_0 + \mathcal{O}(|t|h^\infty\|\varphi_0\|),$$

uniformly with respect to h, t and φ_0.

Reasoning in the same way with $1 - \Pi_g$ instead of Π_g, we also obtain,

$$(1 - \Pi_g)\varphi = e^{-it\tilde{P}^{(2)}/h}(1 - \Pi_g)\varphi_0 + \mathcal{O}(|t|h^\infty\|\varphi_0\|),$$

and (7.2) follows.

2) Formula (7.4) follows exactly in the same way.

3) Since $\Pi_g - \tilde{\Pi}_0 = \mathcal{O}(h)$, for h small enough we can consider the operator $\mathcal{V}$ defined by the Nagy formula,

$$\mathcal{V} = \left(\tilde{\Pi}_0\Pi_g + (1 - \tilde{\Pi}_0)(1 - \Pi_g)\right)\left(1 - (\Pi_g - \tilde{\Pi}_0)^2\right)^{-1/2}. \tag{7.6}$$

Then, $\mathcal{V}$ is a twisted h-admissible operator, it differs from the identity by $\mathcal{O}(h)$, and standard computations (using that $(\Pi_g - \tilde{\Pi}_0)^2$ commutes with both $\tilde{\Pi}_0\Pi_g$ and $(1 - \tilde{\Pi}_0)(1 - \Pi_g)$: see, e.g., [**Ka**] Chap.I.4) show that,

$$\mathcal{V}^*\mathcal{V} = \mathcal{V}\mathcal{V}^* = 1 \quad \text{and} \quad \tilde{\Pi}_0\mathcal{V} = \mathcal{V}\Pi_g.$$

Now, with $\tilde{u}_k$ as in Lemma 3.1, we define $Z_L : L^2(\mathbb{R}^n; \mathcal{H}) \to L^2(\mathbb{R}^n)^{\oplus L}$ by,

$$Z_L\psi(x) = \bigoplus_{k=L'+1}^{L'+L} \langle\psi(x), \tilde{u}_k(x)\rangle_{\mathcal{H}},$$

and we set,

$$\mathcal{W} := Z_L \circ \mathcal{V} = Z_L + \mathcal{O}(h). \tag{7.7}$$

Thanks to the properties of $\mathcal{V}$, we see that $\mathcal{W}\Pi_g = \mathcal{W}$, and, since $Z_L^*Z_L = \tilde{\Pi}_0$ and $Z_LZ_L^* = 1$, we also obtain:

$$\mathcal{W}^*\mathcal{W} = \mathcal{V}^*\tilde{\Pi}_0\mathcal{V} = \Pi_g \ ; \ \mathcal{W}\mathcal{W}^* = 1.$$

Moreover, for any $\varphi_j, \chi_j \in C_d^\infty(\Omega_j)$ such that $\chi_j = 1$ near $\operatorname{Supp}\varphi_j$, and for any $\psi \in L^2(\mathbb{R}^n;\mathcal{H})$, we have,

$$\mathcal{W}U_j^{-1}\varphi_j\psi(x) = \bigoplus_{k=L'+1}^{L'+L} \langle \mathcal{V}_j\psi(x), \tilde{u}_{k,j}(x)\rangle_{\mathcal{H}},$$

with $\mathcal{V}_j := U_j\chi_j\mathcal{V}U_j^{-1}\varphi_j$ and $\tilde{u}_{k,j}(x) := U_j(x)\tilde{u}_k(x) \in C^\infty(\Omega_j,\mathcal{H})$. Therefore, $\mathcal{W}U_j^{-1}\varphi_j$ is an h-admissible operator from $L^2(\mathbb{R}^n;\mathcal{H})$ to $L^2(\mathbb{R}^n)^{\oplus L}$, and the first two properties stated on $\mathcal{W}$ are proved. (Actually, one can easily see that $\mathcal{W}$ also verifies a property analog to the first one in Proposition 4.6, and thus, with an obvious extension of the notion of twisted operator, that $\mathcal{W}$ is, indeed, a twisted h-admissible operator from $L^2(\mathbb{R}^n;\mathcal{H})$ to $L^2(\mathbb{R}^n)^{\oplus L}$.)

Then, defining

$$A := \mathcal{W}\tilde{P}\mathcal{W}^* = \mathcal{W}\tilde{P}^{(1)}\mathcal{W}^*, \tag{7.8}$$

we want to prove that A is an h-admissible operator and study its symbol. We first need the following result:

LEMMA 7.3. *For any $\ell \geq 0$, any $N \geq 1$ and any $\chi_1, \cdots, \chi_N \in C_b^\infty(\mathbb{R}^n)$, one has,*

$$\|\tilde{P}^\ell \mathrm{ad}_{\chi_1} \circ \cdots \circ \mathrm{ad}_{\chi_N}(\mathcal{W}^*)\|_{\mathcal{L}(L^2(\mathbb{R}^n);L^2(\mathbb{R}^n;\mathcal{H})} = \mathcal{O}(h^N). \tag{7.9}$$

Proof – Since $\mathcal{W}^* = \mathcal{V}^*Z_L^*$ and Z_L^* commutes with the multiplication by any function of x, it is enough to prove,

$$\tilde{P}^\ell \mathrm{ad}_{\chi_1} \circ \cdots \circ \mathrm{ad}_{\chi_N}(\mathcal{V}^*) = \mathcal{O}(h^N),$$

on $L^2(\mathbb{R}^n;\mathcal{H})$. Moreover, using (6.3) and and the fact that $\tilde{\Pi}_0$ commutes with the multiplication by any function of x, too, we see on (7.6) that it is enough to show that,

$$(\tilde{P}+i)^\ell(1-(\Pi_g-\tilde{\Pi}_0)^2)^{-1/2}(\tilde{P}+i)^{-\ell} = \mathcal{O}(1); \tag{7.10}$$

$$\tilde{P}^\ell \mathrm{ad}_{\chi_1} \circ \cdots \circ \mathrm{ad}_{\chi_N}\left((1-(\Pi_g-\tilde{\Pi}_0)^2)^{-1/2}\right) = \mathcal{O}(h^N). \tag{7.11}$$

By construction, we have $\tilde{P}^\ell(\Pi(g) - g(\tilde{P})\tilde{\Pi}_0) = \mathcal{O}(h)$, and thus, we immediately see on (6.29) that $\tilde{P}^\ell(\tilde{\Pi}_g - \tilde{\Pi}_0) = \mathcal{O}(h)$. Then, writing

$$\Pi_g - \tilde{\Pi}_0 = \frac{1}{2i\pi}\int_{|z-1|=\frac{1}{2}} (\tilde{\Pi}_g - z)^{-1}(\tilde{\Pi}_0 - \tilde{\Pi}_g)(\tilde{\Pi}_0 - z)^{-1}\, dz,$$

and using Lemma 6.3, we also obtain,

$$\tilde{P}^\ell(\Pi_g - \tilde{\Pi}_0) = \mathcal{O}(h), \tag{7.12}$$

for all $\ell \geq 0$. In particular, $(\tilde{P}+i)^\ell(\Pi_g - \tilde{\Pi}_0)(\tilde{P}+i)^{-\ell} = \mathcal{O}(h)$, and therefore, for h sufficiently small, we can write,

$$(\tilde{P}+i)^\ell(1-(\Pi_g-\tilde{\Pi}_0)^2)^{-1/2}(\tilde{P}+i)^{-\ell} = \left(1 - [(\tilde{P}+i)^\ell(\Pi_g-\tilde{\Pi}_0)(\tilde{P}+i)^{-\ell}]^2\right)^{-1/2},$$

and (7.10) follows.

To prove (7.11), we write $(1 - (\Pi_g - \tilde{\Pi}_0)^2)^{-1/2}$ as,

$$(1 - (\Pi_g - \tilde{\Pi}_0)^2)^{-1/2} = 1 + \sum_{k=1}^{\infty} \alpha_k (\Pi_g - \tilde{\Pi}_0)^k,$$

where the radius of convergence of the power series $\sum_{k=1}^{\infty} \alpha_k z^k$ is 1. Thus,

$$\tilde{P}^\ell \mathrm{ad}_{\chi_1} \circ \cdots \circ \mathrm{ad}_{\chi_N} \left((1 - (\Pi_g - \tilde{\Pi}_0)^2)^{-1/2} \right) = \sum_{k=1}^{\infty} \alpha_k \mathcal{A}_{N,k}$$

where $\mathcal{A}_{N,k} := \tilde{P}^\ell \mathrm{ad}_{\chi_1} \circ \cdots \circ \mathrm{ad}_{\chi_N} ((\Pi_g - \tilde{\Pi}_0)^k)$ is the sum of k^N terms of the form,

$$\tilde{P}^\ell [\mathrm{ad}_{\chi_{i_{1,1}}} \cdots \mathrm{ad}_{\chi_{i_{1,n_1}}} (\Pi_g - \tilde{\Pi}_0)] \ldots [\mathrm{ad}_{\chi_{i_{k,1}}} \cdots \mathrm{ad}_{\chi_{i_{k,n_k}}} (\Pi_g - \tilde{\Pi}_0)],$$

with $n_1, \ldots, n_k \geq 0$, $n_1 + \cdots + n_k = N$. Then, using (6.21) together with (7.12), we see that all these terms have a norm bounded by $(C_N)^k h^{k+N}$, for some constant $C_N > 0$ independent of k. Therefore, $\|\mathcal{A}_{N,k}\| \leq k^N (C_N)^k h^{k+N}$, and (7.11) follows.
●

Then, proceeding as in the proof of Lemma 4.11, we deduce from Lemma 7.9 that, if $\chi, \psi \in C_b^\infty(\mathbb{R}^n)$ are such that dist(Supp χ, Supp ψ) > 0, then, $\|\tilde{P}^\ell \chi \mathcal{W}^* \psi\| = \mathcal{O}(h^\infty)$. As a consequence, taking a partition of unity $(\chi_j)_{j=0,\ldots,r}$ on $\mathbb{R}^n$ with $\chi_j \in C_d^\infty(\Omega_j)$, and choosing $\varphi_j \in C_d^\infty(\Omega_j)$ such that dist(Supp ($\varphi_j - 1$), Supp χ_j) > 0 ($j = 0, \ldots, r$), we have (using also that $\tilde{P}$ is local in the variable x),

$$A = \sum_{j=0}^{r} \mathcal{W} \chi_j \tilde{P} \mathcal{W}^* = \sum_{j=0}^{r} \varphi_j \mathcal{W} \chi_j \tilde{P} \varphi_j^2 \mathcal{W}^* \varphi_j + R(h),$$

with $\|R(h)\|_{\mathcal{L}(L^2(\mathbb{R}^n))} = \mathcal{O}(h^\infty)$. Thus,

$$A = \sum_{j=0}^{r} \varphi_j \mathcal{W} U_j^{-1} \chi_j \tilde{P}_j U_j \varphi_j \mathcal{W}^* \varphi_j + R(h),$$

where $\tilde{P}_j = U_j \tilde{P} U_j^{-1} \varphi_j$ is an h-admissible (differential) operator from $H^m(\mathbb{R}^n; \mathcal{D}_Q)$ to $L^2(\mathbb{R}^n; \mathcal{H})$, while $U_j \varphi_j \mathcal{W}^* \varphi_j$ is an h-admissible operator from $H^m(\mathbb{R}^n)^{\oplus L}$ to $H^m(\mathbb{R}^n; \mathcal{D}_Q)$, and $\varphi_j \mathcal{W} U_j^{-1} \chi_j$ is an h-admissible operator from $L^2(\mathbb{R}^n; \mathcal{H})$ to $L^2(\mathbb{R}^n)^{\oplus L}$.

Therefore, A is an h-admissible operator from $H^m(\mathbb{R}^n)^{\oplus L}$ to $L^2(\mathbb{R}^n)^{\oplus L}$, and, if we set,

$$\tilde{p}_j(x, \xi; h) = \omega(x, \xi; h) + \tilde{Q}_j(x) + \zeta(x) W(x) + h \sum_{|\beta| \leq m-1} \omega_{\beta, j}(x; h) \xi^\beta,$$

and if we denote by $v_j(x, \xi)$ (resp. $v_j^*(x, \xi)$) the symbol of $U_j \mathcal{V} U_j^{-1}$) (resp. $U_j \mathcal{V} U_j^{-1}$), then, the (matrix) symbol $a = (a_{k,\ell})_{1 \leq k, \ell \leq L}$ of A, is given by,

$$a_{k,\ell}(x, \xi, h) = \sum_{j=0}^{r} \langle \chi_j(x) v_j(x, \xi) \sharp \tilde{p}_j(x, \xi) \sharp v_j^*(x, \xi) \sharp \tilde{u}_{L'+k, j}(x), \tilde{u}_{L'+\ell, j}(x) \rangle_{\mathcal{H}}.$$

In particular, since $\partial^\alpha(v_j - 1)$ and $\partial^\alpha(v_j^* - 1)$ are $\mathcal{O}(h)$, we obtain,

$$a_{k,\ell}(x,\xi,h) = \sum_{j=0}^{r} \langle \chi_j(x)(\omega(x,\xi) + \tilde{Q}_j(x) + \zeta(x)W(x))\tilde{u}_{L'+k,j}(x), \tilde{u}_{L'+\ell,j}(x)\rangle_{\mathcal{H}}$$
$$+r_{k,\ell}(h)$$

with $\partial^\alpha r_{k,\ell}(h) = \mathcal{O}(h\langle\xi\rangle^{m-1})$, and thus, using the fact that

$$\langle \tilde{Q}_j(x)\tilde{u}_{L'+k,j}(x), \tilde{u}_{L'+\ell,j}(x)\rangle = \varphi_j(x)\langle \tilde{Q}(x)\tilde{u}_{L'+k}(x), \tilde{u}_{L'+\ell}(x)\rangle,$$

this finally gives,

$$\begin{aligned} a_{k,\ell}(x,\xi,h) &= \sum_{j=0}^{r} \chi_j(x)(\omega(x,\xi)\delta_{k,\ell} + m_{k,\ell}(x) + \zeta(x)W(x)\delta_{k,\ell}) + r_{k,\ell}(h) \\ &= (\omega(x,\xi) + \zeta(x)W(x))\delta_{k,\ell} + m_{k,\ell}(x) + r_{k,\ell}(h), \end{aligned}$$

with $m_{k,\ell}(x) := \langle \tilde{Q}(x)\tilde{u}_{L'+k}(x), \tilde{u}_{L'+\ell}(x)\rangle$. This completes the proof of Theorem 7.1. •

CHAPTER 8

Proof of Theorem 2.1

In view of Theorem 7.1, it is enough to prove,

THEOREM 8.1. *Let* $\varphi_0 \in L^2(\mathbb{R}^n; \mathcal{H})$ *such that* $\|\varphi_0\| = 1$, *and,*

$$\|\varphi_0\|_{L^2(K_0^c;\mathcal{H})} + \|(1-\Pi_g)\varphi_0\| + \|(1-f(P))\varphi_0\| = \mathcal{O}(h^\infty), \tag{8.1}$$

for some $K_0 \subset\subset \Omega' \subset\subset \Omega$, $f, g \in C_0^\infty(\mathbb{R})$, $gf = f$, *and let* $\tilde{P}$ *be the operator constructed in Chapter 2 with* $K = \overline{\Omega'}$, *and* Π_g *be the projection constructed in Theorem 6.1. Then, with the notations of Theorem 7.1, we have,*

$$e^{-itP/h}\varphi_0 = \mathcal{W}^* e^{-itA/h}\mathcal{W}\varphi_0 + \mathcal{O}\left(\langle t\rangle h^\infty\right), \tag{8.2}$$

uniformly with respect to $h > 0$ *small enough and* $t \in [0, T_{\Omega'}(\varphi_0))$.

Proof : Denote by $\chi \in C_0^\infty(\Omega'_K)$ (where Ω'_K is the same as in Proposition 3.2) a cutoff function such that $\chi = 1$ on K. We first prove,

LEMMA 8.2.

$$\|(f(P) - f(\tilde{P}))\chi\|_{\mathcal{L}(L^2(\mathbb{R}^n;\mathcal{H})} = \mathcal{O}(h^\infty).$$

Proof – Using (4.8), we obtain,

$$(f(P) - f(\tilde{P}))\chi = \frac{1}{\pi}\int \overline{\partial}\tilde{f}(z)(P-z)^{-1}(\tilde{P}-P)(\tilde{P}-z)^{-1}\chi dz\, d\bar{z}.$$

Moreover, if $\psi \in C_0^\infty(\Omega'_K)$ is such that $\psi = 1$ on a neighborhood of $\operatorname{Supp}\chi$, Corollary 4.15 and Lemma 4.11 tell us,

$$(\psi - 1)(\tilde{P}-z)^{-1}\chi = \mathcal{O}(h^N|\operatorname{Im} z|^{-(N+1)}),$$

for any $N \geq 1$. As a consequence,

$$(f(P) - f(\tilde{P}))\chi = \frac{1}{\pi}\int \overline{\partial}\tilde{f}(z)(P-z)^{-1}(\tilde{P}-P)\psi(\tilde{P}-z)^{-1}\chi dz\, d\bar{z} + \mathcal{O}(h^\infty),$$

and since $(\tilde{P}-P)\psi = (\tilde{Q}-Q)\psi = 0$, the result follows. •

Now, by (8.1), we have,

$$\varphi_0 = f(P)\varphi_0 + \mathcal{O}(h^\infty) = f(P)\chi\varphi_0 + \mathcal{O}(h^\infty),$$

and thus, by Lemma 8.2,

$$\varphi_0 = f(\tilde{P})\chi\varphi_0 + \mathcal{O}(h^\infty) = f(\tilde{P})\varphi_0 + \mathcal{O}(h^\infty).$$

This means that (7.3) is satisfied, and thus, by Theorem 7.1, the decomposition (7.4) is true. Using (8.1) again, this gives,

$$e^{-it\tilde{P}/h}\varphi_0 = e^{-it\tilde{P}^{(1)}/h}\Pi_g\varphi_0 + \mathcal{O}(|t|h^\infty) = \mathcal{W}^* e^{-itA/h}\mathcal{W}\varphi_0 + \mathcal{O}(\langle t\rangle h^\infty), \tag{8.3}$$

uniformly with respect to h and t.

On the other hand, if we set $\varphi(t) := e^{-itP/h}\varphi_0$, then, by assumption, $\varphi(t) = f(P)\varphi(t) + \mathcal{O}(h^\infty)$ and $\varphi(t) = \chi\varphi(t) + \mathcal{O}(h^\infty)$ uniformly for $t \in [0, T_{\Omega'}(\varphi_0)]$. Therefore, applying Lemma 8.2 again, we obtain as before, $\varphi(t) = f(\tilde{P})\varphi(t) + \mathcal{O}(h^\infty)$, and thus also,

$$\varphi(t) = f(\tilde{P})\chi\varphi(t) + \mathcal{O}(h^\infty), \tag{8.4}$$

uniformly with respect to h and $t \in [0, T_{\Omega'}(\varphi_0)]$. Moreover, since P and $\tilde{P}$ coincide on the support of χ, we can write,

$$ih\partial_t f(\tilde{P})\chi\varphi(t) = f(\tilde{P})\chi P\varphi(t) = f(\tilde{P})\tilde{P}\chi\varphi(t) + f(\tilde{P})[\chi, \tilde{P}]\varphi(t),$$

and thus, since $f(\tilde{P})[\chi, \tilde{P}] = f(\tilde{P})[\chi, \boldsymbol{\omega}]$ is bounded, and $[\chi, \boldsymbol{\omega}]$ is a differential operator with coefficients supported in $\operatorname{Supp}\nabla\chi$ (where φ is $\mathcal{O}(h^\infty)$), we obtain,

$$ih\partial_t f(\tilde{P})\chi\varphi(t) = f(\tilde{P})\chi P\varphi(t) = \tilde{P}f(\tilde{P})\chi\varphi(t) + \mathcal{O}(h^\infty).$$

As a consequence,

$$f(\tilde{P})\chi\varphi(t) = e^{-it\tilde{P}/h}f(\tilde{P})\chi\varphi_0 + \mathcal{O}(|t|h^\infty),$$

and therefore, by (8.4),

$$\varphi(t) = e^{-it\tilde{P}/h}\varphi_0 + \mathcal{O}(\langle t\rangle h^\infty), \tag{8.5}$$

uniformly with respect to h and $t \in [0, T_{\Omega'}(\varphi_0))$. Then, Theorem 8.1 follows from (8.3) and (8.5). •

CHAPTER 9

Proof of Corollary 2.6

First of all, let us recall the (standard) notion of frequency set $FS(v)$ of some (possibly h-dependent) $v \in L^2_{\rm loc}(\Omega)$ (see, e.g., [**Ma2**] and references therein). It is said that a point $(x_0, \xi_0) \in T^*\Omega$ is not in $FS(v)$ if there exist $\chi_1 \in C_0^\infty(\omega)$ and $\chi_2 \in C_0^\infty(\mathbb{R}^n)$ such that $\chi_1(x_0) = \chi_2(\xi_0) = 1$ and $\|\chi_2(hD_x)\chi_1 v\|_{L^2(\mathbb{R}^n)} = \mathcal{O}(h^\infty)$. This is also equivalent to say that there exists an open neighborhood $\mathcal{N}$ of (x_0, ξ_0) in $T^*\mathbb{R}^n$, such that, for *any* $\chi \in C_0^\infty(\mathcal{N})$ and any $\chi_1 \in C_0^\infty(\Omega)$, one has $\|\mathrm{Op}_h(\chi)\chi_1 v\|_{L^2(\mathbb{R}^n)} = \mathcal{O}(h^\infty)$.

As one can see, this notion can be extended in an obvious way to functions in $L^2_{\rm loc}(\Omega; \mathcal{H})$, and it is easy to see (e.g., as in [**Ma2**] Section 2.9) that the latter property still holds with operator-valued functions $\chi \in C_0^\infty(\mathcal{N}; \mathcal{L}(\mathcal{H}))$, or even more generally, $\chi \in C_0^\infty(\mathcal{N}; \mathcal{L}(\mathcal{H}; \mathcal{H}'))$ where $\mathcal{H}'$ is an arbitrary Hilbert-space.

We first prove,

LEMMA 9.1. *Let $\mathcal{W} : L^2(\mathbb{R}^n; \mathcal{H}) \to L^2(\mathbb{R}^n)$ be the operator given in Theorem 7.1. Then, for any $j \in \{0, 1, \dots, r\}$, any $\varphi \in L^2(\mathbb{R}^n; \mathcal{H})$ and $v \in L^2(\mathbb{R}^n)$, such that $\|\varphi\| = \|v\| = 1$, one has,*

$$
\begin{aligned}
FS(\mathcal{W}\varphi) \cap T^*\Omega_j &= FS(U_j \Pi_g \varphi) \cap T^*\Omega_j; \\
FS(U_j \mathcal{W}^* v) \cap T^*\Omega_j &= FS(v) \cap T^*\Omega_j.
\end{aligned}
$$

Proof – Since $\mathcal{W}\mathcal{W}^* = 1$ and $\mathcal{W}^*\mathcal{W} = \Pi_g$, it is enough to prove the two inclusions $FS(\mathcal{W}\varphi) \cap T^*\Omega_j \subset FS(U_j \Pi_g \varphi) \cap T^*\Omega_j$ and $FS(U_j \mathcal{W}^* v) \cap T^*\Omega_j \subset FS(v) \cap T^*\Omega_j$.

Therefore, let $(x_0, \xi_0) \in T^*\Omega_j$, and assume first that $(x_0, \xi_0) \notin FS(U_j \Pi_g \varphi)$. In particular, this implies that, if $\mathcal{N} \subset\subset T^*\Omega_j$ is a small enough neighborhood of (x_0, ξ_0), then $\|\mathrm{Op}_h(\chi_1) U_j \Pi_g \varphi\| = \mathcal{O}(h^\infty)$ for all $\chi_1 \in C_0^\infty(\mathcal{N}; \mathcal{L}(\mathcal{H}; \mathbb{C}))$. Then, taking $\chi \in C_0^\infty(\mathcal{N})$ and $\psi_j \in C_0^\infty(\Omega_j)$ such that $\psi_j(x) = 1$ near $\pi_x(\,\mathrm{Supp}\,\chi)$ and $\chi(x_0, \xi_0) = 1$, we write,

$$
\begin{aligned}
\mathrm{Op}_h(\chi)\mathcal{W}\varphi &= \mathrm{Op}_h(\chi)\mathcal{W}\Pi_g\varphi = \mathrm{Op}_h(\chi)\mathcal{W}\psi_j^2\Pi_g\varphi + \mathcal{O}(h^\infty) \\
&= \mathrm{Op}_h(\chi)\mathcal{W}U_j^{-1}\psi_j U_j \psi_j \Pi_g \varphi + \mathcal{O}(h^\infty),
\end{aligned}
$$

and since $\mathrm{Op}_h(\chi)\mathcal{W}U_j^{-1}\psi_j$ is an h-admissible operator from $L^2(\mathbb{R}^n; \mathcal{H})$ to $L^2(\mathbb{R}^n)$, with symbol supported in $\mathcal{N}$ (that is, modulo $\mathcal{O}(h^\infty)$ in $C_b^\infty(\mathbb{R}^n; \mathcal{L}(\mathcal{H}; \mathbb{C})))$, we obtain $\|\mathrm{Op}_h(\chi)\mathcal{W}\varphi\| = \mathcal{O}(h^\infty)$, and thus $(x_0, \xi_0) \notin FS(\mathcal{W}\varphi)$.

Now, assume that $(x_0, \xi_0) \notin FS(v)$. Since $U_j \psi_j \mathcal{W}^*$ is an h-admissible operator, we obtain in the same way that $\|\mathrm{Op}_h(\chi) U_j \psi_j \mathcal{W}^* v\| = \mathcal{O}(h^\infty)$, and thus $(x_0, \xi_0) \notin FS(U_j \mathcal{W}^* v)$. •

Without loss of generality, we can assume $T_{\Omega'}(\varphi_0) < +\infty$. By Theorem 8.1, we have,

$$e^{-itP/h}\varphi_0 = \mathcal{W}^* e^{-itA/h}\mathcal{W}\varphi_0 + \mathcal{O}(h^\infty),$$

uniformly for $t \in [0, T_{\Omega'}(\varphi_0)]$, where $\mathcal{W}$ and A are given in Theorem 7.1. Thus, by Lemma 9.1, we immediately obtain,

$$FS(U_j e^{-itP/h}\varphi_0) \cap T^*\Omega_j = FS(e^{-itA/h}\mathcal{W}\varphi_0) \cap T^*\Omega_j.$$

On the other hand, since A is an h-admissible operator on $L^2(\mathbb{R}^n)$, a well-known result of propagation (see, e.g., [**Ma2**] Section 4.6, Exercise 12) tells us,

$$FS(e^{-itA/h}\mathcal{W}\varphi_0) = \exp tH_{a_0}(FS(\mathcal{W}\varphi_0)).$$

Therefore, applying Lemma 9.1 again, we obtain,

$$FS(U_j e^{-itP/h}\varphi_0) \cap T^*\Omega_j = T^*\Omega_j \cap \exp tH_{a_0}\left(\cup_{k=0}^r FS(U_k \Pi_g \varphi_0) \cap T^*\Omega_k\right). \tag{9.1}$$

By assumption, we also have,

$$\cup_{k=0}^r FS(U_k\Pi_g\varphi_0) = \cup_{k=1}^r FS(U_k\varphi_0) \subset K_0 \times \mathbb{R}^n. \tag{9.2}$$

In order to conclude, we need the following result:

LEMMA 9.2. *For any $f \in C_0^\infty(\mathbb{R})$, $\psi \in C_0^\infty(\mathbb{R}^n)$, $\chi_j \in C_0^\infty(\Omega_j)$, $\varepsilon > 0$, and $\rho \in C_b^\infty(\mathbb{R})$ with* $\operatorname{Supp}\rho \subset [C_f - \gamma + \varepsilon, +\infty)$ *(where C_f is as in Corollary 2.6), one has,*

$$\|\rho(\chi_j\boldsymbol{\omega}\chi_j)\psi f(U_j\chi_j\tilde{P}U_j^{-1}\chi_j)\| = \mathcal{O}(h^\infty).$$

Proof – We set $\boldsymbol{\omega}_j := \chi_j\boldsymbol{\omega}\chi_j$ and $\tilde{P}_j := U_j\chi_j\tilde{P}U_j^{-1}\chi_j$. Using Assumptions (H1), (H2), (H4) and Proposition 3.2, we see that $\tilde{P}_j \geq (1-Ch)\boldsymbol{\omega}_j + \gamma - Ch$ for some constant $C > 0$ independent of h. As a consequence, we have,

$$\rho(\boldsymbol{\omega}_j)\tilde{P}_j\rho(\boldsymbol{\omega}_j) \geq \rho(\boldsymbol{\omega}_j)((1-Ch)\boldsymbol{\omega}_j + \gamma - Ch)\rho(\boldsymbol{\omega}_j) \geq (C_f + \varepsilon - C'h)\rho(\boldsymbol{\omega}_j)^2,$$

with $C' = C + CC_f$. Therefore, we can write,

$$\|\rho(\boldsymbol{\omega}_j)\psi f(\tilde{P}_j)u\|^2 \leq \frac{1}{C_f+\varepsilon-C'h}\langle \tilde{P}_j\rho(\boldsymbol{\omega}_j)\psi f(\tilde{P}_j)u, \rho(\boldsymbol{\omega}_j)\psi f(\tilde{P}_j)u\rangle,$$

for any $u \in L^2(\mathbb{R}^n;\mathcal{H})$, and thus,

$$\begin{aligned}\|\rho(\boldsymbol{\omega}_j)\psi f(\tilde{P}_j)\| &\leq \frac{1}{C_f+\varepsilon-C'h}\|\tilde{P}_j\rho(\boldsymbol{\omega}_j)\psi f(\tilde{P}_j)\| \\ &\leq \frac{1}{C_f+\varepsilon-C'h}\left(\|\rho(\boldsymbol{\omega}_j)\psi\tilde{P}_j f(\tilde{P}_j)\| + \|[\tilde{P}_j, \rho(\boldsymbol{\omega}_j)\psi]f(\tilde{P}_j)\|\right).\end{aligned} \tag{9.3}$$

Now, on the one hand, since $\operatorname{Supp} f$ is included in $[-C_f, C_f]$, we have,

$$\begin{aligned}\frac{1}{C_f+\varepsilon-C'h}\|\rho(\boldsymbol{\omega}_j)\psi\tilde{P}_j f(\tilde{P}_j)\| &= \frac{1}{C_f+\varepsilon-C'h}\|\tilde{P}_j f(\tilde{P}_j)\psi\rho(\boldsymbol{\omega}_j)\| \\ &\leq \frac{C_f}{C_f+\varepsilon-C'h}\|f(\tilde{P}_j)\psi\rho(\boldsymbol{\omega}_j)\|.\end{aligned} \tag{9.4}$$

On the other hand, since $\tilde{P}_j$ and $\boldsymbol{\omega}_j$ are both differential operators with respect to x with smooth (operator-valued) coefficients, and $\rho(\boldsymbol{\omega}_j)\psi$ is a scalar operator, by standard symbolic calculus, we have,

$$[\tilde{P}_j, \rho(\boldsymbol{\omega}_j)\psi]f(\tilde{P}_j) = \mathcal{O}(h)\rho_1(\boldsymbol{\omega}_j)\psi_1 f(\tilde{P}_j) + \mathcal{O}(h^\infty), \tag{9.5}$$

where $\rho_1 \in C_b^\infty(I\!R)$ and $\psi_1 \in C_0^\infty(I\!R^n)$ are arbitrary functions verifying $\rho_1\rho = \rho$ and $\psi_1\psi = \psi$. Inserting (9.4)-(9.5) into (9.3), we obtain,

$$\|\rho(\boldsymbol{\omega}_j)\psi f(\tilde{P}_j)\| = \mathcal{O}(h\|\rho_1(\boldsymbol{\omega}_j)\psi_1 f(\tilde{P}_j)\|) + \mathcal{O}(h^\infty).$$

Iterating the procedure, we clearly obtain the lemma. •

Now, using, e.g., (8.4), we know that $e^{-itP/h}\varphi_0 = f(\tilde{P})e^{-itP/h}\varphi_0 + \mathcal{O}(h^\infty)$. Moreover, if $\chi_j, \psi_j \in C_0^\infty(\Omega_j)$ are such that $\chi_j = 1$ near Supp ψ_j, by Lemma 4.11, we have,

$$U_j\psi_j f(\tilde{P}) = U_j\psi_j f(\tilde{P})\chi_j^2 + \mathcal{O}(h^\infty) = U_j\psi_j f(\tilde{P})U_j^{-1}\chi_j U_j\chi_j + \mathcal{O}(h^\infty),$$

and therefore,

$$U_j\psi_j e^{-itP/h}\varphi_0 = U_j\psi_j f(\tilde{P})U_j^{-1}\chi_j U_j\chi_j e^{-itP/h}\varphi_0 + \mathcal{O}(h^\infty).$$

Then, using lemma C.1, we obtain,

$$U_j\psi_j e^{-itP/h}\varphi_0 = \psi_j f(\tilde{P}_j)U_j\chi_j e^{-itP/h}\varphi_0 + \mathcal{O}(h^\infty),$$

with $\tilde{P}_j = U_j\chi_j\tilde{P}U_j^{-1}\chi_j$. Therefore, using Lemma 9.2, this gives,

$$\|\rho(\chi_j\boldsymbol{\omega}\chi_j)U_j\psi_j e^{-itP/h}\varphi_0\| = \mathcal{O}(h^\infty),$$

and thus, by Lemma C.2,

$$\|\rho(\boldsymbol{\omega})U_j\psi_j e^{-itP/h}\varphi_0\| = \mathcal{O}(h^\infty). \tag{9.6}$$

Since the principal symbol of $\rho(\boldsymbol{\omega})$ is $\rho(\omega)$, we deduce from (9.2), (9.6), and standard results on FS, that,

$$\cup_{k=0}^r FS(U_k\Pi_g\varphi_0) \subset K(f) := \{(x,\xi)\,;\, x \in K_0\,,\, \omega(x,\xi) \le C_f - \gamma\},$$

and thus, by (9.1),

$$FS(U_j e^{-itP/h}\varphi_0) \cap T^*\Omega_j \subset \exp tH_{a_0}(K(f)) \cap T^*\Omega_j, \tag{9.7}$$

for all $t \ge 0$.

Then, for any $j \in \{0,1,\dots,r\}$, $\psi_j, \tilde{\psi}_j \in C_0^\infty(\Omega_j)$ with $\tilde{\psi}_j\psi_j = \psi_j$, and any $\alpha \in C_0^\infty(I\!R^n)$, we write,

$$U_j\psi_j e^{-itP/h}\varphi_0 = \alpha(hD_x)\tilde{\psi}_j(x)U_j\psi_j e^{-itP/h}\varphi_0 + (1-\alpha(hD_x))U_j\psi_j e^{-itP/h}\varphi_0,$$

and therefore, if $\alpha(\xi) = 1$ in a sufficiently large compact set,

$$U_j\psi_j e^{-itP/h}\varphi_0 = \alpha(hD_x)\tilde{\psi}_j(x)U_j\psi_j e^{-itP/h}\varphi_0 + \mathcal{O}(h^\infty).$$

Finally, if Supp $\tilde{\psi}_j \cap \pi_x(\exp tH_{a_0}(K(f))) = \emptyset$ (or, more generally, Supp $\tilde{\psi}_j \cap \pi_x(\cup_{k=0}^r \exp tH_{a_0}(FS(U_k\Pi_g\varphi_0))) = \emptyset$), then, (9.1) and (9.7) tell us,

$$\|\alpha(hD_x)\tilde{\psi}_j(x)U_j\psi_j e^{-itP/h}\varphi_0\| = \mathcal{O}(h^\infty),$$

and thus, by the unitarity of U_j,

$$\|\psi_j e^{-itP/h}\varphi_0\| = \|U_j\psi_j e^{-itP/h}\varphi_0\| = \mathcal{O}(h^\infty),$$

uniformly for $t \in [0, T_{\Omega'}(\varphi_0)]$. Since we also know that $\|e^{-itP/h}\varphi_0\|_{K^c} = \mathcal{O}(h^\infty)$ for some compact set $K \subset I\!R^n$ (by definition of $T_{\Omega'}(\varphi_0)$), this proves that we can actually take for K any compact neighborhood of $\pi_x(\exp tH_{a_0}(K(f)))$. Thus, if $T_{\Omega'}(\varphi_0) < \sup\{T > 0\,;\, \pi_x(\cup_{t\in[0,T]}\exp tH_{a_0}(K(f))) \subset \Omega'\}$, clearly (e.g., by using Theorem B.1), one can find $T > T_{\Omega'}(\varphi_0)$ and $K_T \subset\subset \Omega'$, such that

$$\sup_{t\in[0,T]} \|e^{-itP/h}\varphi_0\|_{K_T^c} = \mathcal{O}(h^\infty).$$

This is in contradiction with the definition of $T_{\Omega'}(\varphi_0)$, and therefore, necessarily,

$$T_{\Omega'}(\varphi_0) \geq \sup\{T > 0\,;\, \pi_x(\cup_{t\in[0,T]} \exp tH_{a_0}(K(f))) \subset \Omega'\}.$$

This proves Corollary 2.6, and also Remark 2.8 since, in the last argument, one can replace $K(f)$ by $\cup_{k=0}^{r} \exp tH_{a_0}(FS(U_k\Pi_g\varphi_0))$ everywhere. •

CHAPTER 10

Computing the Effective Hamiltonian

Now that we know the existence of an effective Hamiltonian describing the evolution of those states φ_0 that verify (2.4), the problem remains of computing its symbol up to any arbitrary power of h (in Theorem 2.1, only the principal symbol of A is given). Because of the conditions of localization (2.4), it is clear that such an effective Hamiltonian is not unique (for instance, the three operators A, $Af(A)$ or $\mathcal{W}f(\tilde{P})\mathcal{W}^*A\mathcal{W}f(\tilde{P})\mathcal{W}^*$ could indifferently be taken). However, its symbol is certainly uniquely determined in the relevant region of the phase space where $\tilde{\varphi}(t) := \mathcal{W}e^{-itP/h}\varphi_0$ lives (that is, on $FS(\tilde{\varphi}(t))$ in the sense of the previous chapter, and for $t \in [0, T_{\Omega'}(\varphi_0)))$. Therefore, as long as we deal with h-admissible operators (that is, with operators that do not move the Frequency Set), or even with *twisted* h-admissible operators (that become standard h-admissible operators once conjugated with $\mathcal{W}$ or Z_L) it is enough, for computing the symbol A in this region, to start by performing formal computations on the operators themselves (instead of immediately using the twisted symbolic calculus, that appears to be a little bit too heavy at the beginning).

In this chapter, we describe a rather easy way to perform these computations, and we give a simple expression of the effective Hamiltonian up to $\mathcal{O}(h^4)$. Moreover, as an example, we also compute its symbol, up to $\mathcal{O}(h^3)$, in the case $L = 1$. Let us inform the reader that the results of this chapter are not used in the rest of the paper (except for Theorem 12.3), and thus can be skipped without problem at a first reading.

We start from the definition of A given in Chapter 7 (in particular (7.8)):

$$A = \mathcal{W}\tilde{P}\mathcal{W}^* = Z_L\mathcal{V}\tilde{P}\mathcal{V}^*Z_L^*.$$

Since Z_L is rather explicit, the problem mainly consists in determining the expansion of $\mathcal{V}$. Setting,

$$\Delta := h^{-1}(\Pi_g - \tilde{\Pi}_0),$$

and using that $\Pi_g^2 - \Pi_g = \tilde{\Pi}_0^2 - \tilde{\Pi}_0 = 0$, we immediately obtain,

$$\Pi_g\Delta + \Delta\Pi_g = \Delta + h\Delta^2. \tag{10.1}$$

Thus, we deduce from (7.6),

$$\begin{aligned}\mathcal{V} &= ((\Pi_g - h\Delta)\Pi_g + (1 - \Pi_g + h\Delta)(1 - \Pi_g))(1 - h^2\Delta^2)^{-\frac{1}{2}} \\ &= (1 + h[\Pi_g, \Delta] - h^2\Delta^2)(1 - h^2\Delta^2)^{-\frac{1}{2}}.\end{aligned}$$

Then, using the (convergent) series expansion,

$$(1 - h^2\Delta^2)^{-\frac{1}{2}} = 1 + \sum_{k=1}^{\infty} \nu_k h^{2k}\Delta^{2k},$$

with,

$$\nu_k = \frac{1}{2}(\frac{1}{2}+1)(\frac{1}{2}+2)\dots(\frac{1}{2}+k-1)\frac{1}{k!} = \frac{(2k-1)!}{2^{2k-1}k!(k-1)!},$$

we obtain,

$$\mathcal{V} = 1 - ih\mathcal{V}_1 + h^2\mathcal{V}_2,$$

where the two selfadjoint operators $\mathcal{V}_1$ and $\mathcal{V}_2$ are given by,

$$\begin{aligned}\mathcal{V}_1 &= i[\Pi_g, \Delta](1+\sum_{k=1}^{\infty}\nu_k h^{2k}\Delta^{2k});\\ \mathcal{V}_2 &= -\frac{1}{2}\Delta^2 + \sum_{k=1}^{\infty}(\nu_{k+1}-\nu_k)h^{2k}\Delta^{2(k+1)},\end{aligned}$$

that is, observing that $\nu_k - \nu_{k+1} = \nu_k/(2k+2)$,

$$\begin{aligned}\mathcal{V}_1 &= i[\Pi_g, \Delta]F_1(\Delta^2);\\ \mathcal{V}_2 &= F_2(\Delta^2),\end{aligned}$$

with, (setting also $\nu_0 := 1$),

$$\begin{aligned}F_1(s) &= \sum_{k=0}^{\infty}\nu_k h^{2k}s^k;\\ F_2(s) &= -\sum_{k=0}^{\infty}\frac{\nu_k}{2(k+1)}h^{2k}s^{k+1}.\end{aligned}$$

As a consequence,

$$\mathcal{V}^* = 1 + ih\mathcal{V}_1 + h^2\mathcal{V}_2,$$

and therefore,

$$\begin{aligned}\mathcal{V}\tilde{P}\mathcal{V}^* = \tilde{P} + ih[\tilde{P},\mathcal{V}_1] + h^2(\mathcal{V}_1\tilde{P}\mathcal{V}_1 + \mathcal{V}_2\tilde{P} + \tilde{P}\mathcal{V}_2) + ih^3(\mathcal{V}_2\tilde{P}\mathcal{V}_1 - \mathcal{V}_1\tilde{P}\mathcal{V}_2)\\ +h^4\mathcal{V}_2\tilde{P}\mathcal{V}_2,\end{aligned}$$

that is,

$$\begin{aligned}A = Z_L(\tilde{P} + ih[\tilde{P},\mathcal{V}_1] + h^2(\mathcal{V}_1\tilde{P}\mathcal{V}_1 + \mathcal{V}_2\tilde{P} + \tilde{P}\mathcal{V}_2) + ih^3(\mathcal{V}_2\tilde{P}\mathcal{V}_1 - \mathcal{V}_1\tilde{P}\mathcal{V}_2)\\ +h^4\mathcal{V}_2\tilde{P}\mathcal{V}_2)Z_L^*.\end{aligned}$$

From now on, we work modulo $\mathcal{O}(h^5)$ error-terms, and, as we observed at the beginning of this chapter, if we restrict our attention to the relevant region of the phase space, then formal computations are sufficient and Π_g can be replaced by the formal series $\tilde{\Pi} := \sum_{k\geq 0} h^k\tilde{\Pi}_k$ constructed in Chapter 6. In particular, $\tilde{P}$ formally commutes with $\tilde{\Pi}$ and thus, since $[\tilde{P}, \tilde{\Pi}_0] = -ihS_0$ (see Chapter 6),

$$[\tilde{P},[\tilde{\Pi},\Delta]] = -h^{-1}[\tilde{P},[\tilde{\Pi},\tilde{\Pi}_0]] = -h^{-1}[\tilde{\Pi},[\tilde{P},\tilde{\Pi}_0]] = i[\tilde{\Pi},S_0], \tag{10.2}$$

where, from now on, Δ stands for $h^{-1}(\tilde{\Pi}-\tilde{\Pi}_0) = \sum_{k\geq 1} h^k\tilde{\Pi}_k$.

Moreover, from the identities $[\tilde{P},\tilde{\Pi}] = 0$, $\tilde{\Pi} = \tilde{\Pi}_0 + h\Delta$, we deduce,

$$[\tilde{P},\Delta] = -h^{-1}[\tilde{P},\tilde{\Pi}_0] = iS_0,$$

and therefore,

$$\begin{aligned}[\tilde{P},\mathcal{V}_1] &= [S_0,\tilde{\Pi}]F_1(\Delta^2)+i[\tilde{\Pi},\Delta][\tilde{P},F_1(\Delta^2)];\\ [\tilde{P},F_1(\Delta^2)] &= i\sum_{k=1}^{\infty}\nu_k h^{2k}\sum_{j=0}^{2k-1}\Delta^j S_0\Delta^{2k-1-j}.\end{aligned}$$

Since $\nu_0=1$ and $\nu_1=1/2$, this gives,

$$[\tilde{P},\mathcal{V}_1]=[S_0,\tilde{\Pi}](1+\frac{h^2}{2}\Delta^2)-\frac{h^2}{2}[\tilde{\Pi},\Delta](S_0\Delta+\Delta S_0)+\mathcal{O}(h^4) \tag{10.3}$$

Moreover, (10.1) implies $\tilde{\Pi}\Delta\tilde{\Pi}=h\Delta^2\tilde{\Pi}=h\tilde{\Pi}\Delta^2$, and thus, in particular, Δ^2 commutes with $\tilde{\Pi}$. As a consequence, we can write,

$$\begin{aligned}\mathcal{V}_1\tilde{P}\mathcal{V}_1 &= F_1(\Delta^2)[\tilde{\Pi},\Delta]\tilde{P}[\Delta,\tilde{\Pi}]F_1(\Delta^2)\\ &= [\tilde{\Pi},\Delta]\tilde{P}[\Delta,\tilde{\Pi}]+h^2\ \mathrm{Re}\ \Delta^2[\tilde{\Pi},\Delta]\tilde{P}[\Delta,\tilde{\Pi}]+\mathcal{O}(h^4),\end{aligned}$$

and, still using (10.1), we have,

$$\begin{aligned}[\tilde{\Pi},\Delta]\tilde{P}[\Delta,\tilde{\Pi}] &= \tilde{\Pi}\Delta\tilde{P}\Delta\tilde{\Pi}+\Delta\tilde{\Pi}\tilde{P}\tilde{\Pi}\Delta-\tilde{\Pi}\Delta\tilde{P}\tilde{\Pi}\Delta-\Delta\tilde{\Pi}\tilde{P}\Delta\tilde{\Pi}\\ &= (\tilde{\Pi}\Delta+\Delta\tilde{\Pi})\tilde{P}(\Delta\tilde{\Pi}+\tilde{\Pi}\Delta)-2\tilde{\Pi}\Delta\tilde{P}\tilde{\Pi}\Delta-2\Delta\tilde{\Pi}\tilde{P}\Delta\tilde{\Pi}\\ &= (\Delta+h\Delta^2)\tilde{P}(\Delta+h\Delta^2)-2h\tilde{\Pi}\Delta^2\tilde{P}\Delta-2h\Delta\tilde{P}\Delta^2\tilde{\Pi}\\ &= \Delta\tilde{P}\Delta+h(1-2\tilde{\Pi})\Delta^2\tilde{P}\Delta+h\Delta\tilde{P}\Delta^2(1-2\tilde{\Pi})\\ &= \frac{1}{2}(\Delta^2P+P\Delta^2)+\frac{i}{2}[\Delta,S_0]+2h\ \mathrm{Re}\ \Delta^2(1-2\tilde{\Pi})\tilde{P}\Delta.\end{aligned}$$

Therefore,

$$\begin{aligned}\mathcal{V}_1\tilde{P}\mathcal{V}_1= &\ \mathrm{Re}\ \Delta^2P+\frac{i}{2}[\Delta,S_0]+2h\ \mathrm{Re}\ \Delta^2(1-2\tilde{\Pi})\tilde{P}\Delta\\ &+h^2\ \mathrm{Re}\ \Delta^2(\ \mathrm{Re}\ \Delta^2P+\frac{i}{2}[\Delta,S_0])+\mathcal{O}(h^3).\end{aligned}$$

and, since $\mathcal{V}_2=-\frac{1}{2}\Delta^2-\frac{1}{8}h^2\Delta^4+\mathcal{O}(h^4)$, we obtain,

$$\begin{aligned}\mathcal{V}_1\tilde{P}\mathcal{V}_1+\mathcal{V}_2\tilde{P}+\tilde{P}\mathcal{V}_2 &= \frac{i}{2}[\Delta,S_0]+2h\ \mathrm{Re}\ \Delta^2(1-2\tilde{\Pi})\tilde{P}\Delta\\ &\quad+h^2\left(\ \mathrm{Re}\ \Delta^2(\ \mathrm{Re}\ \Delta^2P+\frac{i}{2}[\Delta,S_0])-\frac{1}{4}\ \mathrm{Re}\ \Delta^4\tilde{P}\right)\\ &\quad+\mathcal{O}(h^3)\\ &= \frac{i}{2}[\Delta,S_0]+2h\ \mathrm{Re}\ \Delta^2(1-2\tilde{\Pi})\tilde{P}\Delta\\ &\quad+\frac{1}{2}h^2\left(\ \mathrm{Re}\ (i\Delta^2[\Delta,S_0])+\Delta^2\tilde{P}\Delta^2+\frac{1}{4}\ \mathrm{Re}\ \Delta^4\tilde{P}\right)\\ &\quad+\mathcal{O}(h^3)\end{aligned}$$

Finally, since, obviously, Δ^2 also commutes with Δ, thus with $[\tilde{\Pi},\Delta]$, too, we see that $\mathcal{V}_1$ and $\mathcal{V}_2$ commute together, and therefore,

$$\begin{aligned}\mathcal{V}_2\tilde{P}\mathcal{V}_1-\mathcal{V}_1\tilde{P}\mathcal{V}_2 &= [\tilde{P},\mathcal{V}_1]\mathcal{V}_2-[\tilde{P},\mathcal{V}_2]\mathcal{V}_1\\ &= -\frac{1}{2}[S_0,\tilde{\Pi}]\Delta^2+\frac{i}{2}[\tilde{P},\Delta^2][\tilde{\Pi},\Delta]+\mathcal{O}(h^2)\\ &= -\frac{1}{2}[S_0,\tilde{\Pi}]\Delta^2-\frac{1}{2}(S_0\Delta+\Delta S_0)[\tilde{\Pi},\Delta]+\mathcal{O}(h^2).\end{aligned}$$

Summing up, we have found,

$$\mathcal{V}\tilde{P}\mathcal{V}^* = B_0 + hB_1 + h^2B_2 + h^3B_3 + h^4B_4 + \mathcal{O}(h^5),$$

with,

$$\begin{aligned}
B_0 &= \tilde{P}\\
B_1 &= i[S_0, \tilde{\Pi}]\\
B_2 &= \frac{i}{2}[\Delta, S_0]\\
B_3 &= -\text{ Re } i[\tilde{\Pi}, \Delta](S_0\Delta + \Delta S_0) + 2 \text{ Re } \Delta^2(1-2\tilde{\Pi})\tilde{P}\Delta\\
B_4 &= \frac{1}{2}\Big(\text{ Re } (i\Delta^2[\Delta, S_0]) + \Delta^2\tilde{P}\Delta^2 + \frac{1}{4}\text{ Re } \Delta^4\tilde{P}\Big)
\end{aligned}$$

Then, writing $\tilde{\Pi} = \sum_{k=0}^{3} h^k\tilde{\Pi}_k + \mathcal{O}(h^4)$ and $\Delta = \sum_{k=1}^{3} h^{k-1}\tilde{\Pi}_k + \mathcal{O}(h^3)$, we obtain,

$$\mathcal{V}\tilde{P}\mathcal{V}^* = C_0 + hC_1 + h^2C_2 + h^3C_3 + h^4C_4 + \mathcal{O}(h^5),$$

with,

$$\begin{aligned}
C_0 &= \tilde{P}\\
C_1 &= i[S_0, \tilde{\Pi}_0]\\
C_2 &= \frac{i}{2}[S_0, \tilde{\Pi}_1]\\
C_3 &= \frac{i}{2}[S_0, \tilde{\Pi}_2] - \text{ Re } i[\tilde{\Pi}_0, \tilde{\Pi}_1](S_0\tilde{\Pi}_1 + \tilde{\Pi}_1 S_0) + 2 \text{ Re } \tilde{\Pi}_1^2(1-2\tilde{\Pi}_0)\tilde{P}\tilde{\Pi}_1\\
C_4 &= \frac{i}{2}[S_0, \tilde{\Pi}_3] - \text{ Re } i[\tilde{\Pi}_0, \tilde{\Pi}_2](S_0\tilde{\Pi}_1 + \tilde{\Pi}_1 S_0) - \text{ Re } i[\tilde{\Pi}_0, \tilde{\Pi}_1](S_0\tilde{\Pi}_2 + \tilde{\Pi}_2 S_0)\\
&\quad +2 \text{ Re } (\tilde{\Pi}_1\tilde{\Pi}_2 + \tilde{\Pi}_2\tilde{\Pi}_1)(1-2\tilde{\Pi}_0)\tilde{P}\tilde{\Pi}_1 - 4 \text{ Re } \tilde{\Pi}_1^3\tilde{P}\tilde{\Pi}_1\\
&\quad +2 \text{ Re } \tilde{\Pi}_1^2(1-2\tilde{\Pi}_0)\tilde{P}\tilde{\Pi}_2 + \frac{1}{2}\Big(\text{ Re } (i\tilde{\Pi}_1^2[\tilde{\Pi}_1, S_0]) + \tilde{\Pi}_1^2\tilde{P}\tilde{\Pi}_1^2 + \frac{1}{4}\text{ Re } \tilde{\Pi}_1^4\tilde{P}\Big)
\end{aligned}$$

Now, due to (6.7)-(6.8), we observe that,

$$\tilde{\Pi}_0 S_0 \tilde{\Pi}_0 = \tilde{\Pi}_0^\perp S_0 \tilde{\Pi}_0^\perp = \tilde{\Pi}_0\tilde{\Pi}_1\tilde{\Pi}_0 = \tilde{\Pi}_0^\perp\tilde{\Pi}_1\tilde{\Pi}_0^\perp = 0.$$

As a consequence,

$$\tilde{\Pi}_0 C_1 \tilde{\Pi}_0 = i\tilde{\Pi}_0[S_0, \tilde{\Pi}_0]\tilde{\Pi}_0 = 0,$$

and,

$$\tilde{\Pi}_0[\tilde{\Pi}_0, \tilde{\Pi}_1](S_0\tilde{\Pi}_1 + \tilde{\Pi}_1 S_0)\tilde{\Pi}_0 = \tilde{\Pi}_0[\tilde{\Pi}_0, \tilde{\Pi}_1]\tilde{\Pi}_0(S_0\tilde{\Pi}_1 + \tilde{\Pi}_1 S_0)\tilde{\Pi}_0 = 0;$$

$$\begin{aligned}
\tilde{\Pi}_0\tilde{\Pi}_1^2(1-2\tilde{\Pi}_0)\tilde{P}\tilde{\Pi}_1\tilde{\Pi}_0 &= \tilde{\Pi}_0\tilde{\Pi}_1^2\tilde{\Pi}_0^\perp\tilde{P}\tilde{\Pi}_1\tilde{\Pi}_0 + \tilde{\Pi}_0\tilde{\Pi}_1^2(1-2\tilde{\Pi}_0)[\tilde{P}, \tilde{\Pi}_0^\perp]\tilde{\Pi}_1\tilde{\Pi}_0\\
&= ih\tilde{\Pi}_0\tilde{\Pi}_1^2(1-2\tilde{\Pi}_0)S_0\tilde{\Pi}_1\tilde{\Pi}_0\\
&= -ih\tilde{\Pi}_0\tilde{\Pi}_1^2 S_0\tilde{\Pi}_1\tilde{\Pi}_0.
\end{aligned}$$

(In the last two steps we have used that $\tilde{\Pi}_0\tilde{\Pi}_1^2\tilde{\Pi}_0^\perp = \tilde{\Pi}_0^\perp S_0\tilde{\Pi}_1\tilde{\Pi}_0 = 0$.) Since we also have $Z_L = Z_L\tilde{\Pi}_0$ and $Z_L^* = \tilde{\Pi}_0 Z_L^*$, we deduce,

$$Z_L C_1 Z_L^* = 0;$$

(10.4) $$Z_L C_3 Z_L^* = \frac{i}{2} Z_L[S_0, \tilde{\Pi}_2]Z_L^* + 2h \text{ Im } \tilde{\Pi}_0\tilde{\Pi}_1^2 S_0\tilde{\Pi}_1\tilde{\Pi}_0.$$

In particular, since $A = Z_L\mathcal{V}\tilde{P}\mathcal{V}^* Z_L^*$, we have proved,

PROPOSITION 10.1. *The effective Hamiltonian A verifies,*

$$A = A_0 + h^2 A_2 + h^3 A_3 + \mathcal{O}(h^4), \tag{10.5}$$

with,

$$\begin{aligned} A_0 &= Z_L \tilde{P} Z_L^* \\ A_2 &= \frac{i}{2} Z_L [S_0, \tilde{\Pi}_1] Z_L^* \\ A_3 &= \frac{i}{2} Z_L [S_0, \tilde{\Pi}_2] Z_L^*. \end{aligned}$$

It is interesting to observe that, at this level, the absence of a term in h (that is, an extra-term of the form hA_1) is completely general and, in particular, is not related to any particular form of $\boldsymbol{\omega}$ (however, some term in h may be hidden in A_0, as we shall see in the sequels).

Here, we have stopped the computation of A at the third power of h, but it is clear from the expression of C_4 and (10.4) that the coefficient of h^4 can be written down, too (but has a more complicated form). Of course, pushing forward the series and spending more time in the calculation would permit to also obtain the next terms.

From that point, in order to have an even more explicit expression of A (in particular to compute its symbol), one must use the expressions of $\tilde{\Pi}_1$ and $\tilde{\Pi}_2$ obtained in Chapter 6. Let us do it in the case $L = 1$. In that case, setting $\lambda(x) := \lambda_{L'+1}(x)$, one has $\tilde{\Pi}_0(z - \tilde{Q}(x))^{-1} = (z - \lambda(x))^{-1}\tilde{\Pi}_0$, and thus,

$$\tilde{\Pi}_0^{\perp}\tilde{\Pi}_1\tilde{\Pi}_0 = -\frac{1}{2\pi}\oint_{\gamma(x)} \frac{(z - \tilde{Q}(x))^{-1}\tilde{\Pi}_0^{\perp}(x) S_0 \tilde{\Pi}_0(x)}{z - \lambda(x)} dz = -iR'(\lambda(x))S_0,$$

where $R'(x, z) := \tilde{\Pi}_0^{\perp}(x)(z - \tilde{Q}(x))^{-1}\tilde{\Pi}_0^{\perp}(x)$ is the so-called *reduced resolvent* of $\tilde{Q}(x)$.

As a consequence,

$$\tilde{\Pi}_0[S_0, \tilde{\Pi}_1]\tilde{\Pi}_0 = S_0\tilde{\Pi}_0^{\perp}\tilde{\Pi}_1\tilde{\Pi}_0 - \tilde{\Pi}_0\tilde{\Pi}_1\tilde{\Pi}_0^{\perp} S_0 = -2iS_0 R'(x, \lambda(x)) S_0,$$

that leads to,

$$A_2 = Z_1 S_0 R'(x, \lambda(x)) S_0 Z_1^*.$$

In the same way,

$$\tilde{\Pi}_0^{\perp}\tilde{\Pi}_2\tilde{\Pi}_0 = -iR'(x, \lambda(x)) S_1 \tilde{\Pi}_0,$$

and therefore,

$$A_3 = \operatorname{Re} Z_1 S_0 R'(x, \lambda(x)) S_1 Z_1^*.$$

Now, we can start to use the twisted symbolic calculus introduced in Chapter 4. We denote by $s_0 = (s_0^j)_{0 \le j \le r}$ and $\pi_0 = (\pi_0^j)_{0 \le j \le r}$ the (twisted) symbols of S_0 and $\tilde{\Pi}_0$ respectively. We also set $\tilde{\omega} = (\tilde{\omega}_j)_{0 \le j \le r}$, where,

$$\tilde{\omega}_j(x, \xi) := \omega(x, \xi) + h \sum_{|\beta| \le m-1} \omega_{\beta,j}(x)\xi^\beta, \quad ((x.\xi) \in T^*\Omega_j),$$

is the symbol of the operator introduced in (2.3) (we remind that we work with the standard quantization of symbols, as described in Chapter A). From (6.4)-(6.6) and the considerations of Chapter 4 (and since $\pi_0^j = \pi_0^j(x)$ does not depend on ξ), it s easy to see that,

$$s_0^j = \partial_\xi \tilde{\omega}_j \partial_x \pi_0^j + i \sum_{|\beta|\leq m-1} [\omega_{\beta,j}(x), \pi_0^j(x)]\xi^\beta - \frac{ih}{2} \sum_{|\alpha|=2} (\partial_\xi^\alpha \omega)(\partial_x^\alpha \pi_0^j) + \mathcal{O}(h^2)$$

If we also set,

$$\tilde{Q}_j(x) := U_j(x)\tilde{Q}(x)U_j(x)^{-1},$$

then, the symbol $\rho = (\rho_j)_{0\leq j\leq r}$ of $R'(x, \lambda(x))$ is simply given by,

$$\rho_j(x) = (1 - \pi_0^j(x))(\lambda(x) - \tilde{Q}_j(x))^{-1}(1 - \pi_0^j(x)),$$

and thus, the symbol $\sigma_2 = (\sigma_2^j)_{0\leq j\leq r}$ of $S_0 R'(x, \lambda(x))S_0$ verifies,

$$\sigma_2^j(x,\xi) = s_0^j(x,\xi)\rho_j(x)s_0^j(x,\xi) + \frac{h}{i}\partial_\xi s_0^j(x,\xi)\partial_x(\rho_j(x)s_0^j(x,\xi)) + \mathcal{O}(h^2).$$

From (6.8)-(6.10), we also obtain,

$$\begin{aligned} \tilde{\Pi}_1 &= i[S_0, R'(x, \lambda(x))] \\ S_1 &= \frac{i}{h}[\omega + \zeta W, \tilde{\Pi}_1]. \end{aligned}$$

Therefore, since ω and ζW are scalar operators, the respective symbols $\pi_1 = (\pi_1^j)_{0\leq j\leq r}$ and $s_1 = (s_1^j)_{0\leq j\leq r}$ of $\tilde{\Pi}_1$ and S_1, verify,

$$\begin{aligned} \pi_1^j(x,\xi) &= i[s_0^j(x,\xi), \rho_j(x)] + \mathcal{O}(h) = i\partial_\xi\omega(x,\xi)[\partial_x\pi_0^j(x), \rho_j(x)] + \mathcal{O}(h) \\ s_1^j &= \{\omega + \zeta W, \pi_1^j\} + \mathcal{O}(h) = \partial_\xi\omega \cdot \partial_x \pi_1^j - \partial_\xi \pi_1^j \cdot \partial_x(\omega + \zeta W) + \mathcal{O}(h), \end{aligned}$$

and thus,

$$s_1^j = i \sum_{k,\ell=1}^{n} \Big((\partial_{\xi_k}\omega)\partial_{x_k}(\partial_{\xi_\ell}\omega[\partial_{x_\ell}\pi_0^j, \rho_j]) - (\partial_{\xi_k}\partial_{\xi_\ell}\omega)[\partial_{x_\ell}\pi_0^j, \rho_j]\partial_{x_k}(\omega + \zeta W) \Big) + \mathcal{O}(h). \tag{10.6}$$

This permits to compute the symbol $\sigma_3 = (\sigma_3^j)_{0\leq j\leq r}$ of $\operatorname{Re} S_0 R'(x, \lambda(x))S_1$, by using the formula,

$$\sigma_3^j(x,\xi) = \frac{1}{2}\partial_\xi\omega \cdot \Big((\partial_x\pi_0^j)\rho_j s_1^j + s_1^j \rho_j (\partial_x \pi_0^j) \Big) + \mathcal{O}(h). \tag{10.7}$$

Observe that one also has,

$$\partial_x \pi_0^j(x) = \langle \cdot, \nabla_x u_j(x)\rangle_{\mathcal{H}} u_j(x) + \langle \cdot, u_j(x)\rangle_{\mathcal{H}} \nabla_x u_j(x),$$

where $\langle \cdot, u\rangle_{\mathcal{H}}$ stands for the operator $w \mapsto \langle w, u\rangle_{\mathcal{H}}$, and $u_j =: U_j(x)u_{L'+1}(x)$ is the normalized eigenfunction of $\tilde{Q}_j(x)$ associated with $\lambda(x)$.

Finally, we use the following elementary remark: let B is a twisted h-admissible (or PDO) operator on $L^2(\mathbb{R}^n; \mathcal{H})$, with symbol $b = (b_j)_{0\leq j\leq r}$, and let $u(x), v(x) \in \mathcal{H}$ such that, for all $j = 0, \dots, r$, $u_j(x) := U_j(x)u(x)$ and $v_j(x) := U_j(x)v(x)$ are in $C^\infty(\Omega_j; \mathcal{H})$. Denote by $\mathcal{Z}_u$, $\mathcal{Z}_v$ the operators $L^2(\mathbb{R}^n; \mathcal{H}) \to L^2(\mathbb{R}^n)$ defined by ,

$$\mathcal{Z}_u w := \langle w, u\rangle_{\mathcal{H}} \ ; \ \mathcal{Z}_v w := \langle w, v\rangle_{\mathcal{H}}.$$

Then, the symbol $\check{b}$ of the (standard) h-admissible operator $\mathcal{Z}_v B \mathcal{Z}_u^*$ verifies,

$$\forall (x,\xi) \in T^*\Omega_j, \ \check{b}(x,\xi) = \langle b_j(x,\xi)\sharp u_j(x), v_j(x)\rangle_{\mathcal{H}},$$

where the operation $\sharp$ is defined in an obvious way, by substituting the usual product with the action of an operator (here, the various derivatives of $b_j(x,\xi)$) on a function (here, the various derivatives of $u_j(x)$).

We can clearly apply this remark to compute the symbol of A_2 and A_3, but also that of A_0, since we have,

$$A_0 = Z_1\tilde{P}Z_1^* = \mathcal{Z}_u\tilde{P}\mathcal{Z}_u^* = \mathcal{Z}_{Q_0u}Q_0^{-1}\tilde{P}\mathcal{Z}_u^*,$$

with $u := \tilde{u}_{L'+1}$ (defined in Chapter 3), and, by Proposition 5.5, we know that $Q_0^{-1}\tilde{P}$ is a twisted PDO.

Combining all the previous computations, using that $\tilde{Q}_j(x)u_j(x) = \lambda(x)u_j(x)$ for all $j = 0, \dots, r$ and $x \in \Omega_j$, and gathering (as far as possible) the terms with same homogeneity in h, we finally arrive to the following result (leaving some details to the reader):

PROPOSITION 10.2. *In the case* $\mathrm{Rank}\,\Pi_0(x) = 1$, *the effective Hamiltonian* A *verifies (10.5) with,*

$$
\begin{aligned}
A_0 &= Z_1\tilde{P}Z_1^*;\\
A_2 &= \frac{1}{h^2}Z_1[\tilde{P},\tilde{\Pi}_0]R'(x,\lambda(x))[\tilde{\Pi}_0,\tilde{P}]Z_1^*; \qquad (10.8)\\
A_3 &= \frac{1}{h^3}\,\mathrm{Re}\, Z_1[\tilde{P},\tilde{\Pi}_0]R'(x,\lambda(x))[[[\tilde{P},\tilde{\Pi}_0],R'(x,\lambda(x))],\boldsymbol{\omega}+\zeta W]Z_1^*,
\end{aligned}
$$

where $\lambda(x)$ *is the (only) eigenvalue of* $\tilde{Q}(x)\tilde{\Pi}_0$, *and* $R'(x,\lambda(x)) = \tilde{\Pi}_0^{\perp}(x)(\lambda(x)-\tilde{Q}(x))^{-1}\tilde{\Pi}_0^{\perp}(x)$ *is the reduced resolvent of* $\tilde{Q}(x)$.

Moreover, the symbol $a(x,\xi;h)$ *of* A *verifies,*

$$a(x,\xi;h) = a_0(x,\xi) + ha_1(x,\xi) + h^2a_2(x,\xi) + \mathcal{O}(h^3),$$

with, for any $(x,\xi) \in T^*\Omega_j$ ($j = 0,\dots,r$ *arbitrary*),

$$
\begin{aligned}
a_0(x,\xi) &= \omega(x,\xi;h) + \lambda(x) + \zeta(x)W(x);\\
a_1(x,\xi) &= \sum_{|\beta|\le m-1}\langle\omega_{\beta,j}(x)u_j(x),u_j(x)\rangle\xi^\beta - i\langle\nabla_\xi\omega(x,\xi)\nabla_x u_j(x),u_j(x)\rangle;\\
a_2(x,\xi) &= \sum_{k,\ell=1}^{n}(\partial_{\xi_k}\omega)(\partial_{\xi_\ell}\omega)\langle\rho_j(x)\partial_{x_k}u_j,\partial_{x_\ell}u_j\rangle - \frac{1}{2}\sum_{|\alpha|=2}(\partial_\xi^\alpha\omega)\langle\partial_x^\alpha u_j,u_j\rangle\\
&\quad -i\sum_{|\beta|\le m-1}\langle\omega_{\beta,j}(x)\nabla_x u_j(x),u_j(x)\rangle\cdot\nabla_\xi(\xi^\beta)\\
&\quad -2\,\mathrm{Im}\sum_{|\beta|\le m-1}\nabla_\xi\omega(x,\xi)\langle\omega_{\beta,j}(x)\rho_j(x)\nabla_x u_j(x),u_j(x)\rangle\xi^\beta\\
&\quad +\sum_{|\beta|,|\gamma|\le m-1}\langle\omega_{\beta,j}(x)\rho_j(x)u_j(x),\omega_{\beta,j}(x)u_j(x)\rangle\xi^{\beta+\gamma}.
\end{aligned}
$$

REMARK 10.3. *Although some of these terms may seem to depend on the choice of* j *verifying* $(x,\xi)\in T^*\Omega_j$, *actually we know that this cannot be the case. In fact, the independency with respect to* j *is due to the compatibility conditions (4.10) satisfied by the symbols of twisted pseudodifferential operators.*

REMARK 10.4. *Actually, it results from the previous computations that (10.8) is still valid in the (slightly) more general case where* L *is arbitrary and* $\lambda_{L'+1}(x) = \cdots = \lambda_{L'+L}(x)$ *for all* $x \in \Omega$.

REMARK 10.5. *Using (10.6)-(10.7), one can find an expression for the* h^3*-term of the symbol of* A*, too. We leave it as an exercise to the reader.*

CHAPTER 11

Propagation of Wave-Packets

In this chapter, we assume $L = 1$ and we make the following additional assumption on the coefficients c_α of $\boldsymbol{\omega}$:

$$(11.1) \qquad c_\alpha(x;h) \sim \sum_{k=0}^{\infty} h^k c_{\alpha,k}(x),$$

with $c_{\alpha,k}$ independent of h. Then, in a similar spirit as in [**Ha6**], we investigate the evolution of an initial state of the form,

$$(11.2) \qquad \varphi_0(x) = (\pi h)^{-n/4} f(P)\Pi_g(e^{ix\xi_0/h-(x-x_0)^2/2h}u_{L'+1}(x)),$$

where $(x_0,\xi_0) \in T^*\Omega$ is fixed, $f, g \in C_0^\infty(I\!R)$ are such that $f = 1$ near $a_0(x_0,\xi_0)$ (here, $a_0(x,\xi)$ is the same as in Corollary 2.6), $g = 1$ near $\operatorname{Supp} f$, and Π_g is constructed as in Chapter 6, starting from the operator $\tilde{P}$ constructed in Chapter 3 with $K \ni x_0$. In particular, since $e^{-(x-x_0)^2/2h}$ is exponentially small for x outside any neighborhood of x_0, by Lemma 8.2, we have,

$$\varphi_0(x) = (\pi h)^{-n/4} f(\tilde{P})\Pi_g(e^{ix\xi_0/h-(x-x_0)^2/2h}\tilde{u}_{L'+1}(x)) + \mathcal{O}(h^\infty),$$

in $L^2(I\!R^n;\mathcal{H})$. Moreover, due to the properties of Π_g, and the fact that the coherent state $\phi_0 := (\pi h)^{-n/4}e^{ix\xi_0/h-(x-x_0)^2/2h}$ is normalized in $L^2(I\!R^n)$, we also obtain,

$$\varphi_0(x) = (\pi h)^{-n/4} f(\tilde{P})e^{ix\xi_0/h-(x-x_0)^2/2h}\tilde{u}_{L'+1}(x) + \mathcal{O}(h),$$

and thus, in particular, $\|\varphi_0\| = 1 + \mathcal{O}(h)$. Actually, we even have the following better result:

Proposition 11.1. *The function φ_0 admits, in $L^2(I\!R^n;\mathcal{H})$, an asymptotic expansion of the form,*

$$(11.3) \qquad \varphi_0(x) \sim (\pi h)^{-n/4}e^{ix\xi_0/h-(x-x_0)^2/2h}\sum_{k=0}^{\infty} h^k v_k(x) + \mathcal{O}(h^\infty),$$

with $v_k \in L^\infty(I\!R^n;\mathcal{H})$ $(k \geq 0)$, and $v_0(x) = \tilde{u}_{L'+1}(x) + \mathcal{O}(|x-x_0|)$ in $\mathcal{H}$, uniformly with respect to $x \in I\!R^n$. Moreover, for any $j \in \{0,1,\dots,r\}$ and any $\chi_j \in C_d^\infty(\Omega_j)$, the function $U_j\chi_j\varphi_0$ admits, in $C_d^\infty(\Omega_j;\mathcal{H})$, an asymptotic expansion of the form,

$$(11.4) \qquad U_j(x)\chi_j(x)\varphi_0(x) \sim (\pi h)^{-n/4}e^{ix\xi_0/h-(x-x_0)^2/2h}\sum_{k=0}^{\infty} h^k\chi_j(x)v_{j,k}(x) + \mathcal{O}(h^\infty),$$

with $v_{j,k} \in C^\infty(\Omega_j;\mathcal{H})$, $v_{j,0}(x) = U_j(x)\tilde{u}_{L'+1}(x) + \mathcal{O}(|x-x_0|)$.

Proof – For $j = 0, 1, \dots, r$, let $\chi_j \in C_d^\infty(\Omega_j)$, such that $\sum \chi_j = 1$, and let $\tilde{\chi}_j \in C_d^\infty(\Omega_j)$, such that $\tilde{\chi}_j = 1$ near $\operatorname{Supp} \chi_j$. Then, since $f(\tilde{P})$ and Π_g are twisted h-admissible operators, have,

$$\begin{aligned}
\varphi_0 &= \sum_j \chi_j \varphi_0 \\
&= \sum_j U_j^{-1} \tilde{\chi}_j U_j \chi_j f(\tilde{P}) \tilde{\chi}_j^2 \Pi_g \tilde{\chi}_j^2 (\phi_0(x) \tilde{u}_{L'+1}(x)) + \mathcal{O}(h^\infty) \\
&= \sum_j U_j^{-1} \tilde{\chi}_j U_j \chi_j f(\tilde{P}) U_j^{-1} \tilde{\chi}_j U_j \tilde{\chi}_j \Pi_g \tilde{\chi}_j^2 (\phi_0(x) \tilde{u}_{L'+1}(x)) + \mathcal{O}(h^\infty),
\end{aligned}$$

and thus, by Lemma C.1, and setting $\tilde{P}_j := U_j \tilde{\chi}_j \tilde{P} U_j^{-1} \tilde{\chi}_j$, $\Pi_{g,j} := U_j \tilde{\chi}_j \Pi_g U_j^{-1} \tilde{\chi}_j$, and $u_{L'+1,j}(x) := U_j(x) \tilde{\chi}_j(x) \tilde{u}_{L'+1}(x)$ ($\in C_d^\infty(\Omega_j; \mathcal{H})$), we obtain,

$$\varphi_0 = \sum_{j=0}^{r} U_j^{-1} \chi_j f(\tilde{P}_j) \Pi_{g,j} (\phi_0(x) u_{L'+1,j}(x)) + \mathcal{O}(h^\infty). \tag{11.5}$$

Now, using the results of Chapters 4 and 6, we see that $f(\tilde{P}_j)\Pi_{g,j}$ is an h-admissible operator on $L^2(I\!R^n; \mathcal{H})$, with symbol b_j verifying,

$$\begin{aligned}
& b_j(x, \xi; h) \sim \sum_{k=0}^{\infty} h^k b_{j,k}(x, \xi); \\
& b_{j,0}(x, \xi) = f(\tilde{\chi}_j(x)^2 (\omega_0(x, \xi) + \tilde{Q}_j(x) + W(x))) \tilde{\chi}_j(x)^2 \tilde{\Pi}_{0,j}(x),
\end{aligned}$$

where $\omega_0(x, \xi) := \sum_{|\alpha| \leq m} c_{\alpha,0}(x) \xi^\alpha$, $\tilde{Q}_j(x) = U_j(x) \tilde{Q}(x) U_j(x)^{-1}$, and $\tilde{\Pi}_{0,j}(x) = U_j(x) \tilde{\Pi}_0(x) U_j(x)^{-1}$. Moreover, we have,

$$\operatorname{Op}_h(b_j)(\phi_0 u_{L'+1,j})(x; h) = \frac{1}{(2\pi h)^n} \int e^{i(x-y)\xi/h + iy\xi_0/h} \rho(x, y, \xi; h) dy d\xi,$$

with,

$$\rho(x, y, \xi; h) = (\pi h)^{-n/4} e^{-(y-x_0)^2/2h} b_j(x, \xi; h) u_{L'+1,j}(y),$$

and it is easy to check that, for any $\alpha, \beta \in Z\!\!Z_+^n$, one has,

$$\|(hD_y)^\alpha (hD_\xi)^\beta \rho(x, y, \xi; h)\|_{\mathcal{H}} = \mathcal{O}(h^{|\alpha|/2 + |\beta|}),$$

uniformly for $(x, y, \xi) \in I\!R^{3n}$ and $h > 0$ small enough. As a consequence, we can perform a standard stationary phase expansion in the previous (oscillatory) integral (see, e.g., [**DiSj1, Ma2**]), and since the unique critical point is given by $y = x$ and $\xi = \xi_0$, we obtain,

$$\operatorname{Op}_h(b_j)(\phi_0 v_j)(x; h) = e^{ix\xi_0/h} w_j(x; h) + \mathcal{O}(h^\infty),$$

with,

$$w_j(x; h) \sim \sum_{k=0}^{\infty} \frac{h^k}{i^k k!} (\nabla_y \cdot \nabla_\xi)^k \rho(x, y, \xi; h) \Big|_{\substack{y=x \\ \xi=\xi_0}}.$$

Therefore, since $e^{(y-x_0)^2/2h} \nabla_y e^{-(y-x_0)^2/2h} = \nabla_y - \frac{y-x_0}{h}$, and, for any $k \in I\!N$, $|y - x_0|^k e^{-(y-x_0)^2/2h} = \mathcal{O}(h^{k/2})$, we also obtain,

$$\operatorname{Op}_h(b_j)(\phi_0 u_{L'+1,j})(x; h) = (\pi h)^{-n/4} e^{ix\xi_0/h - (x-x_0)^2/2h} \tilde{w}_j(x; h),$$

with,
(11.6)
$$\tilde{w}_j(x,h) = \sum_{k=0}^{N} \frac{h^k}{i^k k!}((\nabla_y - h^{-1}(y-x_0))\cdot\nabla_\xi)^k b_j(x,\xi;h)u_{L'+1,j}(y)\Big|_{\substack{y=x\\ \xi=\xi_0}} + \mathcal{O}(h^{N/2}),$$
for any $N \geq 0$. Then, taking a ressummation of the formal series in $(x-x_0)$ obtained for each degree of homogeneity in h in (11.6), we obtain an asymptotic expansion of $\tilde{w}_j$, of the form,
$$\tilde{w}_j(x,h) \sim \sum_{k=0}^{\infty} h^k \tilde{w}_{j,k}(x).$$
(Alternatively – and equivalently – one could have used instead the stationary phase theorem with complex-valued phase function [**MeSj1**] Theorem 2.3, with the phase $(x-y)\xi + y\xi_0 + i(y-x_0)^2/2$.) In particular, the first coefficient $\tilde{w}_{j,0}(x)$ is obtained as a resummation of the formal series $\sum_{k\geq 0}\frac{i^k}{k!}((y-x_0))\cdot\nabla_\xi)^k b_j(x,\xi;h)u_{L'+1,j}(y)\Big|_{\substack{y=x\\ \xi=\xi_0}}$, and thus,
$$\begin{aligned}
\tilde{w}_{j,0}(x)(x) &= b_j(x,\xi_0;h)u_{L'+1,j}(x) + \mathcal{O}(|x-x_0|)\\
&= f(\tilde{\chi}_j(x)^2(\omega_0(x,\xi_0) + \tilde{Q}_j(x) + W(x)))\tilde{\chi}_j(x)^2 u_{L'+1,j}(x)\\
&\qquad + \mathcal{O}(|x-x_0|)\\
&= f(\tilde{\chi}_j(x)^2(\omega_0(x,\xi_0) + \lambda_{L'+1}(x) + W(x)))\tilde{\chi}_j(x)^2 u_{L'+1,j}(x)\\
&\qquad + \mathcal{O}(|x-x_0|)\\
&= f(\tilde{\chi}_j(x)^2(a_0(x_0,\xi_0))\tilde{\chi}_j(x)^2 u_{L'+1,j}(x) + \mathcal{O}(|x-x_0|).
\end{aligned}$$
Going back to (11.5), this gives an asymptotic expansion for φ_0 of the form (11.3), with,
$$\begin{aligned}
v_0(x) &= \sum_{j=0}^{r} U_j(x)^{-1}\chi_j(x) f(\tilde{\chi}_j(x)^2 a_0(x_0,\xi_0))u_{L'+1,j}(x) + \mathcal{O}(|x-x_0|)\\
&= \sum_{j=0}^{r} U_j(x)^{-1}\chi_j(x) f(a_0(x_0,\xi_0))u_{L'+1,j}(x) + \mathcal{O}(|x-x_0|)\\
&= \sum_{j=0}^{r} U_j(x)^{-1}\chi_j(x) u_{L'+1,j}(x) + \mathcal{O}(|x-x_0|)\\
&= \tilde{u}_{L'+1}(x) + \mathcal{O}(|x-x_0|).
\end{aligned}$$
The asymptotic expansion (11.4) is obtained exactly in the same way. •

As a consequence, we also obtain,

PROPOSITION 11.2. *For any $j \in \{0,1,\dots,r\}$, one has,*
$$FS(U_j\varphi_0) = \{(x_0,\xi_0)\} \cap T^*\Omega_j.$$

Proof – For $\chi_j \in C_d^\infty(\Omega_j)$ fixed, we denote by $w_j(x;h)$ a resummation of the formal series $\sum_{k\geq 0} h^k U_j(x)\chi_j(x)v_{j,k}(x)$ in $C_d^\infty(\Omega_j;\mathcal{H})$, where the $v_{j,k}$'s are those in (11.4). Then, defining,
$$\begin{aligned}
A = A(x,hD_x) &:= (hD_x - \xi_0)^2 + (x-x_0)^2\\
&= (hD_x - \xi_0 + i(x-x_0))\cdot(hD_x - \xi_0 - i(x-x_0)) + nh,
\end{aligned}$$

a straightforward computation gives,

$$A(U_j\varphi_0) = A(\phi_0(x)w_j(x;h)) + \mathcal{O}(h^\infty) = h\phi_0(x)Bw_j(x;h) + \mathcal{O}(h^\infty)$$

with $Bw_j(x;h) := 2i(x-x_0)\cdot\partial_x w_j - ih\partial_x^2 w_j + nw_j$, and thus, by an iteration,

$$A^N(\phi_0(x)w_j(x;h)) = h^N\phi_0(x)B^N w_j + \mathcal{O}(h^\infty),$$

for any $N \geq 1$. In particular, due to the form of B, and since $\|(x-x_0)^\alpha\phi_0\| = \mathcal{O}(1)$ for any $\alpha \in \mathbb{Z}_+^n$ (actually, $\mathcal{O}(h^{|\alpha|/2})$), we obtain,

$$\|A^N(U_j\varphi_0)\|_{L^2(\Omega_j',\mathcal{H})} = \mathcal{O}(h^N),$$

for any $\Omega_j' \subset\subset \Omega_j$. Now, if $(x_1,\xi_1) \in T^*\Omega_j$ is different from (x_0,ξ_0), then A^N is elliptic at (x_1,ξ_1) and thus, given any $\chi \in C_0^\infty(T^*\Omega_j)$ with $\chi(x_1,\xi_1) = 1$, the standard construction of a microlocal parametrix (see, e.g., [**DiSj1**]) gives an uniformly bounded operator A_N' , such that,

$$A_N' \circ A^N = \chi(x,hD_x) + \mathcal{O}(h^\infty).$$

As a consequence, we obtain,

$$\|\chi(x,hD_x)(U_j\varphi_0)\|_{L^2(\Omega_j',\mathcal{H})} = \mathcal{O}(h^N),$$

for all $N \geq 1$. Therefore $(x_1,\xi_1) \notin FS(\phi_0(x)v_j(x))$, and thus, we have proved,

$$FS(U_j\varphi_0) \subset \{(x_0,\xi_0)\} \cap T^*\Omega_j.$$

This means that $FS(U_j\varphi_0)$ consists in at most one point. Conversely, if $x_0 \in \Omega_j$ and $FS(U_j\varphi_0) = \emptyset$, by the ellipticity of A^N as $|\xi| \to \infty$, we would have (see, e.g., [**Ma2**] Prop. 2.9.7),

$$\|U_j\varphi_0\|_{\Omega_j'} = \mathcal{O}(h^\infty),$$

for any $\Omega_j' \subset\subset \Omega_j$. But this contradicts the fact that $\|U_j\varphi_0\|_{\Omega_j'} = \|\varphi_0\|_{\Omega_j'} = 1+\mathcal{O}(h)$ if $x_0 \in \Omega_j'$. •

Now, applying Theorem 2.1 and Corollary 2.6 (or rather Remark 2.8), we obtain,

$$e^{itP/h}\varphi_0 = \mathcal{W}^*e^{-itA/h}\mathcal{W}\varphi_0 + \mathcal{O}(\langle t\rangle h^\infty), \tag{11.7}$$

uniformly for $t \in [0, T_{\Omega'}(x_0,\xi_0))$,where $\Omega' \subset\subset \Omega$ is the same as the one used to define $\tilde{P}$ in Chapter 3, and

$$T_{\Omega'}(x_0,\xi_0) := \sup\{T > 0\,;\, \pi_x(\cup_{t\in[0,T]}\exp tH_{a_0}(x_0,\xi_0)) \subset \Omega'\}. \tag{11.8}$$

Moreover, by Lemma 9.1 and Proposition 11.2, we see that,

$$FS(\mathcal{W}\varphi_0) = \{(x_0,\xi_0)\}. \tag{11.9}$$

Assuming, e.g., that $x_0 \in \Omega_1$, and taking $\chi_1 \in C_0^\infty(\Omega_1)$ such that $\chi_1 = 1$ in a neighborhood of x_0, we also have,

$$\mathcal{W}\varphi_0 = \mathcal{W}\chi_1^2\varphi_0 + \mathcal{O}(h^\infty) = \mathcal{W}U_1^{-1}\chi_1U_1\chi_1\varphi_0 + \mathcal{O}(h^\infty),$$

and therefore, using (11.4), (7.7), and the fact that $\mathcal{W}U_1^{-1}\chi_1$ is an h-admissible operator from $L^2(\mathbb{R}^n;\mathcal{H})$ to $L^2(\mathbb{R}^n)$ (see Theorem 7.1), we obtain as before (by a stationary phase expansion),

$$\mathcal{W}\varphi_0(x;h) \sim (\pi h)^{-n/4}e^{ix\xi_0/h-(x-x_0)^2/2h}\sum_{k=0}^\infty h^k w_k(x) + \mathcal{O}(h^\infty), \tag{11.10}$$

with $w_k \in C_b^\infty(I\!R^n)$, $w_0(x) = \langle \tilde{u}_{L'+1}(x), \tilde{u}_{L'+1}(x)\rangle + \mathcal{O}(|x-x_0|) = 1 + \mathcal{O}(|x-x_0|)$, and where the asymptotic expansion takes place in $C_b^\infty(I\!R^n)$.

This means that $\mathcal{W}\varphi_0$ is a coherent state in $L^2(I\!R^n)$, centered at (x_0,ξ_0), and from this point we can apply all the standard (and less standard) results of semiclassical analysis for scalar operators, in order to compute $e^{-itA/h}\mathcal{W}\varphi_0$ (see, e.g., [**CoRo, Ha1, Ro1, Ro2**] and references therein). In particular, we learn from [**CoRo**] Theorem 3.1 (see also [**Ro2**]), that, for any $N \geq 1$,

$$e^{-itA/h}\mathcal{W}\varphi_0 = e^{i\delta_t/h}\sum_{k=0}^{3(N-1)} c_k(t;h)\Phi_{k,t} + \mathcal{O}(e^{NC_0t}h^{N/2}), \tag{11.11}$$

where $\Phi_{k,t}$ is a (generalized) coherent state centered at $(x_t,\xi_t) := \exp tH_{a_0}(x_0,\xi_0)$, $\delta_t := \int_0^t(\dot{x}_s\xi_s - a_0(x_s,\xi_s))ds + (x_0\xi_0 - x_t\xi_t)/2$, $C_0>0$ is a constant, the coefficients $c_k(t;h)$'s are of the form,

$$c_k(t;h) = \sum_{\ell=0}^{N_k} h^\ell c_{k,\ell}(t), \tag{11.12}$$

with $c_{k,\ell}$ universal polynomial with respect to $(\partial^\gamma a_0(x_t,\xi_t))_{|\gamma|\leq M_k}$, and where the estimate is uniform with respect to (t,h) such that $0\leq t< T_{\Omega'}(x_0,\xi_0)$ and he^{C_0t} remains bounded ($h>0$ small enough). In particular, (11.11) supplies an asymptotic expansion of $e^{-itA/h}\mathcal{W}\varphi_0$ if one restricts to the values of t such that $0\leq t << \ln\frac{1}{h}$.

Now, applying $\mathcal{W}^*$ to (11.11), and observing that $\mathcal{W}^*\Phi_{k,t} = \mathcal{V}^*(\Phi_{k,t}\tilde{u}_{L'+1}) = U_j^{-1}\mathcal{V}_j^*(\Phi_{k,t}u_{L'+1,j})$, where $j=j(t)$ is chosen in such a way that $\exp tH_{a_0}(x_0,\xi_0)\in\Omega_j$, and where $\mathcal{V}_j^* := U_j\mathcal{V}^*U_j^{-1}$ is an h-admissible operator on $L^2(\Omega_j;\mathcal{H})$ (that is, becomes an h-admissible operator on $L^2(I\!R^n;\mathcal{H})$ once sandwiched by cutoff functions supported in Ω_j), we deduce from (11.7),

THEOREM 11.3. *Let φ_0 be as in (11.2), and let $T_{\Omega'}(x_0,\xi_0)$ defined in (11.8). Then, there exists $C>0$ such that, for any $N\geq 1$, one has,*

$$e^{-itP/h}\varphi_0 = e^{i\delta_t/h}\sum_{k=0}^{3(N-1)} c_k(t;h)\Phi_{k,t}U_{j(t)}^{-1}\tilde{v}_{k,j(t)}(x) + \mathcal{O}(h^{N/4}),$$

where $\Phi_{k,t}$ is a coherent state centered at $(x_t,\xi_t) := \exp tH_{a_0}(x_0,\xi_0)$, $j(t)\in\{1,\dots,r\}$ is such that $\exp tH_{a_0}(x_0,\xi_0)\in\Omega_{j(t)}$, $\tilde{v}_{k,j(t)}\in C^\infty(\Omega_{j(t)};\mathcal{H})$, $c_k(t;h)$ is as in (11.12), $\delta_t := \int_0^t(\dot{x}_s\xi_s - a_0(x_s,\xi_s))ds + (x_0\xi_0 - x_t\xi_t)/2$, and where the estimate is uniform with respect to (t,h) such that $h>0$ is small enough and $t\in[0,\min(T_{\Omega'}(x_0,\xi_0), C^{-1}\ln\frac{1}{h}))$.

REMARK 11.4. *Actually, the coherent state $\Phi_{k,t}$ is of the form,*

$$\Phi_{k,t} = c_k(t)f_k(x,\sqrt{h})h^{-n/4}e^{ix\xi_t/h - q_t(x-x_t)/h},$$

*where $c_k(t)$ is a normalizing factor, f_k is polynomial in 2 variables, and q_t is a t-dependent quadratic form with positive-definite real part, that can be explicitly computed by using a classical evolution involving the Hessian of a_0 at (x_t,ξ_t) (see [**CoRo**]). More precisely, one has $q_t(x) = -i\langle\Gamma_t x, x\rangle/2$ with $\Gamma_t = (C_t+iD_t)(A_t+$*

$iB_t)^{-1}$, *where the* $2n \times 2n$ *matrix,*

$$F_t = \begin{pmatrix} A_t & B_t \\ C_t & D_t \end{pmatrix}$$

is, by definition, the solution of the classical problem,

$$\dot{F}_t = J\text{Hess}\, a_0(x_t, \xi_t)F_t \quad ; \quad F(0) = I_{2n}.$$

Here, $J := \begin{pmatrix} 0 & I_n \\ -I_n & 0 \end{pmatrix}$, *and* Hess a_0 *stands for the Hessian of* a_0. *(We are grateful to M. Combescure and D. Robert for having explained to us this construction and the main result of* [**CoRo**].*)*

REMARK 11.5. *As in* [**CoRo**], *one can also consider more general initial states, of the form,*

$$\varphi_0(x) = e^{i(\xi_0 \cdot x - x_0 \cdot hD_x)/h} f\left(\frac{x}{\sqrt{h}}\right),$$

where $f \in \mathcal{S}(I\!R^n)$ *(we refer to* [**CoRo**] *Theorem 3.5 for more details). In the same way, a similar result can also be obtained for oscillating initial states of the form,*

$$\varphi_0(x) = f(x)e^{iS(x)/h},$$

where $f \in C_0^\infty(I\!R^n)$ *and* $S \in C^\infty(I\!R^n; I\!R)$ *(see* [**CoRo**] *Remark 3.9).*

REMARK 11.6. *In principle, all the terms of the asymptotic series can be computed explicitly by an inductive procedure (although, in practical, this task may result harder than expected since the simplifications are sometimes quite tricky). Indeed, all our constructions mainly rely on symbolic pseudodifferential calculus, that provides very explicit inductive formulas.*

CHAPTER 12

Application to Polyatomic Molecules

In this chapter, we apply all the previous results to the particular case of a polyatomic molecule with Coulomb-type interactions, imbedded in an electromagnetic field. Denoting by $x = (x_1, \dots, x_n) \in \mathbb{R}^{3n}$ the position of the n nuclei, and by $y = (y_1, \dots, y_p) \in \mathbb{R}^{3p}$ the position of the p electrons, the corresponding Hamiltonian takes the form,

$$H = \sum_{j=1}^{n} \frac{1}{2M_j}(D_{x_j} - A(x_j))^2 + \sum_{k=1}^{p} \frac{1}{2m_k}(D_{y_k} - A(y_k))^2 + V(x,y), \tag{12.1}$$

where the magnetic potential A is assumed to be in $C_b^\infty(\mathbb{R}^3)$, and where the electric potential V can be written as,

$$V(x,y) = V_{\text{nu}}(x) + V_{\text{el}}(y) + V_{\text{el-nu}}(x,y) + V_{\text{ext}}(x,y) = V_{\text{int}}(x,y) + V_{\text{ext}}(x,y). \tag{12.2}$$

Here, V_{nu} (resp. V_{el}, resp. $V_{\text{el-nu}}$) stands for sum of the nucleus-nucleus (resp. electron-electron, resp. electron-nucleus) interactions, and V_{ext} stands for the external electric potential. Actually, our techniques can be applied to a slightly more general form of Hamiltonian (also allowing, somehow, a strong action of the magnetic field upon the nuclei), namely,

$$H = \sum_{j=1}^{n} \frac{1}{2M_j}(D_{x_j} - a_j A_j(x))^2 + \sum_{k=1}^{p} \frac{1}{2m_k}(D_{y_k} - B_k(x,y))^2 + V(x,y), \tag{12.3}$$

where $A_1, \dots, A_n$ (respectively $B_1, \dots, B_p$) are assumed to be in $C_b^\infty(\mathbb{R}^n; \mathbb{R})$ (respectively $C_b^\infty(\mathbb{R}^{n+p}; \mathbb{R})$), the a_j's are extra parameters, and V is as in (12.2) with,

$$\begin{aligned} &V_{\text{nu}}(x) = \sum_{1\leq j<j'\leq n} \frac{\alpha_{j,j'}}{|x_j - x_{j'}|} \quad ; \quad V_{\text{el}}(y) = \sum_{1\leq k<k'\leq p} \frac{\beta_{k,k'}}{|y_k - y_{k'}|} \ ; \\ &V_{\text{el-nu}}(x,y) = \sum_{\substack{1\leq j\leq n \\ 1\leq k\leq p}} \frac{-\gamma_{j,k}}{|x_j - y_k|} \quad ; \quad V_{ext} \in C_b^\infty(\mathbb{R}^{n+p}; \mathbb{R}), \end{aligned} \tag{12.4}$$

$\alpha_{j,j'}, \beta_{k,k'}, \gamma_{j,k} > 0$ constant. In fact, as in [**KMSW**], more general forms can be allowed for the interaction potentials, e.g., by replacing any function of the type $|z_j - z'_k|^{-1}$ (where the letters z and z' stand for x or y indifferently) by some $V_{j,k}(z_j - z'_k)$, where $V_{j,k}$ is assumed to be Δ-compact on $L^2(\mathbb{R}^3)$ and to verify some estimates on its derivatives (see [**KMSW**] Section 2). In the same way, one could also have admitted singularities of the same kind for the exterior potentials. However, here we keep the form (12.4) since it is more concrete and corresponds to the usual physical situation.

Then, we consider the Born-Oppenheimer limit in the following sense: We set,

$$M_j = h^{-2}b_j\ ;\ a_j = h^{-1}c_j + d_j, \tag{12.5}$$

and we consider the limit $h \to 0_+$ for some fix $b_j, m_k > 0$, $c_j, d_j \in I\!R$. By scaling the time variable, too, the quantum evolution of the molecule is described by the Schrödinger equation,

$$ih\frac{\partial\varphi}{\partial t} = P(h)\varphi,$$

where,

$$P(h) := \sum_{j=1}^{n} \frac{1}{2b_j}(hD_{x_j} - (c_j + hd_j)A_j(x))^2 + \sum_{k=1}^{p} \frac{1}{2m_k}(D_{y_k} - B_k(x,y))^2 + V(x,y). \tag{12.6}$$

In particular, we see that $P(h)$ satisfies to Assumptions (H1) and (H2), with,

$$\omega = \sum_{j=1}^{n} \frac{1}{2b_j}(hD_{x_j} - (c_j + hd_j)A_j(x))^2,$$

$$\omega(x,\xi;h) = \sum_{j=1}^{n} \frac{1}{2b_j}\left[(\xi_j - (c_j + hd_j)A_j(x))^2 + ih(c_j + hd_j)(\partial_{x_j}A_j)(x)\right],$$

$$Q(x) = \sum_{k=1}^{p} \frac{1}{2m_k}(D_{y_k} - B_k(x,y))^2 + V_{\text{el}}(y) + V_{\text{el-nu}}(x,y) + V_{\text{ext}}(x,y),$$

$$W(x) = V_{\text{nu}}(x).$$

Now, following the terminology of [**KMSW**], we denote by

$$\mathcal{C} := \bigcup_{\substack{1\le j,k\le n\\ j\ne k}} \{x = (x_1,\dots,x_n) \in I\!R^{3n}\ ;\ x_j = x_k\}$$

the so-called *collision set* of nuclei, and we make on $Q(x)$ the following gap condition:

($H3'$) There exists a contractible bounded open set $\Omega \subset I\!R^{3n}$ such that $\overline{\Omega} \cap \mathcal{C} = \emptyset$, and, for all $x \in \overline{\Omega}$, the $L'+L$ first values $\lambda_1(x), \dots, \lambda_{L'+L}(x)$, given by the Mini-Max principle for $Q(x)$ on $L^2(I\!R^{3p})$, are discrete eigenvalues of $Q(x)$, and verify,

$$\inf_{x\in\Omega} \text{dist}\,(\sigma(Q(x))\backslash\{\lambda_{L'+1}(x),\dots,\lambda_{L'+L}(x)\}, \{\lambda_{L'+1}(x),\dots,\lambda_{L'+L}(x)\}) > 0.$$

As it is well known (see [**CoSe**]), under these assumptions, the two spectral projections $\Pi_0^-(x)$ and $\Pi_0(x)$ of $Q(x)$, corresponding to $\{\lambda_1(x),\dots,\lambda_{L'}(x)\}$ and $\{\lambda_{L'+1}(x),\dots,\lambda_{L'+L}(x)\}$ respectively, are twice differentiable with respect to $x \in \Omega$. In particular, the whole assumption (H3) is indeed satisfied in that case (and even with a slightly larger open subset of $I\!R^{3n}$).

Now, in order to be able to apply the results of the previous chapters to this molecular Hamiltonian, it remains to construct a family $(\Omega_j, U_j(x))_{1\le j\le r}$ that verifies Assumption (H4). We do it by following [**KMSW**].

More precisely, for any fixed $x_0 = (x_1^0, \dots, x_n^0) \in I\!R^{3n} \backslash \mathcal{C}$, we choose n functions $f_1, \dots, f_n \in C_0^\infty(I\!R^3; I\!R)$, such that,

$$f_j(x_k^0) = \delta_{j,k} \;\; (1 \le j, k \le n),$$

and, for $x \in I\!R^{3n}$, $s \in I\!R^3$, and $y = (y_1, \dots, y_p) \in I\!R^{3p}$, we set,

$$\begin{aligned} F_{x_0}(x, s) &:= s + \sum_{k=1}^{n} (x_k - x_k^0) f_k(s) \in I\!R^3, \\ G_{x_0}(x, y) &:= (F_{x_0}(x, y_1), \dots, F_{x_0}(x, y_p)) \in I\!R^{3p}. \end{aligned}$$

Then, by the implicit function theorem, for x in a sufficiently small neighborhood Ω_{x_0} of x_0, the application $y \mapsto G_{x_0}(x, y)$ is a diffeomorphism of $I\!R^{3p}$, and we have,

$$\begin{aligned} &x_k = F_{x_0}(x, x_k^0), \\ &G_{x_0}(x, y) = y \text{ for } |y| \text{ large enough.} \end{aligned}$$

Now, for $v \in L^2(I\!R^{3p})$ and $x \in \Omega_{x_0}$, we define,

$$U_{x_0}(x) v(y) := |\det d_y G_{x_0}(x, y)|^{\frac{1}{2}} v(G_{x_0}(x, y))|,$$

and we see that $U_{x_0}(x)$ is a unitary operator on $L^2(I\!R^{3p})$ that preserves both $\mathcal{D}_Q = H^2(I\!R^{3p})$ and $C_0^\infty(I\!R^{3p})$. Moreover, denoting by U_{x_0} the operator on $L^2(\Omega_{x_0} \times I\!R^{3p})$ induced by $U_{x_0}(x)$, we have the following identities:

$$\begin{aligned} &U_{x_0} h D_x U_{x_0}^{-1} = h D_x + h J_1(x, y) D_y + h J_2(x, y), \\ &U_{x_0} D_y U_{x_0}^{-1} = J_3(x, y) D_y + J_4(x, y), \\ &U_{x_0} \frac{1}{|y_k - y_k'|} U_{x_0}^{-1} = \frac{1}{|F_{x_0}(x, y_k) - F_{x_0}(x, y_k'),|} \\ &U_{x_0} \frac{1}{|x_j - y_k|} U_{x_0}^{-1} = \frac{1}{|F_{x_0}(x, x_j^0) - F_{x_0}(x, y_k)|}, \end{aligned} \tag{12.7}$$

where the (matrix or operator-valued) functions J_ν's $(1 \le \nu \le 4)$ are all smooth on $\Omega_{x_0} \times I\!R^{3p}$. Indeed, denoting by $\tilde{G}_{x_0}(x, \cdot)$ the inverse diffeomorphism of $G_{x_0}(x, \cdot)$, one finds,

$$\begin{aligned} J_1(x, y) &= ({}^t d_x \tilde{G}_{x_0})(x, y' = G_{x_0}(x, y)), \\ J_2(x, y) &= |\det d_y G_{x_0}(x, y)|^{\frac{1}{2}} D_x \left(|\det d_{y'} \tilde{G}_{x_0}(x, y')|^{\frac{1}{2}} \right) \Big|_{y' = G_{x_0}(x, y))}, \\ J_3(x, y) &= ({}^t d_{y'} \tilde{G}_{x_0})(x, y' = G_{x_0}(x, y)), \\ J_4(x, y) &= |\det d_y G_{x_0}(x, y)|^{\frac{1}{2}} D_{y'} \left(|\det d_{y'} \tilde{G}_{x_0}(x, y')|^{\frac{1}{2}} \right) \Big|_{y' = G_{x_0}(x, y))}. \end{aligned}$$

The key-point in (12.7) is that the (x-dependent) singularity at $y_k = x_j$ has been replaced by the (fix) singularity at $y_k = x_j^0$. Then, as in [**KMSW**], one can easily deduce that the map $x \mapsto U_{x_0} Q(x) U_{x_0}^{-1}$ is in $C^\infty(\Omega_{x_0}; \mathcal{L}(H^2(I\!R^{3p}), L^2(I\!R^{3p}))$. Moreover, so is the map $x \mapsto U_{x_0} \Delta_y U_{x_0}^{-1}$, and we also see that $U_{x_0} \omega U_{x_0}^{-1}$ can be written as in (2.3) (with Ω_{x_0} instead of Ω_j, $m = 2$, and $Q_0 = -\Delta_y + C_0$, $C_0 > 0$ large enough). Indeed, with the notations of (12.7), and setting $\mathcal{J}(x) =$

$(\mathcal{J}_1(x),\dots,\mathcal{J}_n(x)) := J_1(x,y)D_y + J_2(x,y)$, we have,

$$
\begin{aligned}
U_{x_0}\omega U_{x_0}^{-1} &= \sum_{k=1}^{n}\frac{1}{2b_k}(hD_{x_k}+h\mathcal{J}_k(x)-(c_k+hd_k)A_k(x))^2\\
&= \omega + h\sum_{k=1}^{n}\frac{1}{b_k}\mathcal{J}_k(hD_{x_k}-c_kA_k)\\
&\quad + h^2\sum_{k=1}^{n}\frac{1}{2b_k}(\mathcal{J}_k^2 - i(\nabla_x\mathcal{J}_k)-2d_kA_k\mathcal{J}_k).
\end{aligned}
\tag{12.8}
$$

To complete the argument, we just observe that the previous construction can be made around any point x_0 of $\overline{\Omega}$, and since this set is compact, we can cover it by a finite family $\tilde{\Omega}_1,\dots,\tilde{\Omega}_r$ of open sets such that each one corresponds to some Ω_{x_0} as before. Denoting also $U_1(x),\dots,U_r(x)$ the corresponding operators $U_{x_0}(x)$, and setting $\Omega_j=\tilde{\Omega}_j\cap\Omega$, we can conclude that the family $(\Omega_j,U_j(x))_{1\le j\le r}$ verifies (H4) with $\mathcal{H}_\infty = C_0^\infty(\mathbb{R}^{3p})$. As a consequence, we can apply to this model all the results of the previous chapters, and thus, we have proved,

THEOREM 12.1. *Let $P(h)$ be as in (12.6) with V given by (12.2) and (12.4), $A_1,\dots,A_n\in C_b^\infty(\mathbb{R}^n;\mathbb{R})$, and $B_1,\dots,B_p\in C_b^\infty(\mathbb{R}^{n+p};\mathbb{R})$. Assume also (H3'). Then, the conclusions of Theorem 2.1 are valid for $P=P(h)$.*

We also observe that, in this case, we have,

$$
\omega(x,\xi;h)=\omega_0(x,\xi)+h\omega_1(x,\xi)+h^2\omega_2(x),
$$

with,

$$
\begin{aligned}
\omega_0(x,\xi) &= \sum_{k=1}^{n}\frac{1}{2b_k}(\xi_k-c_kA_k(x))^2\\
\omega_1(x,\xi) &= \sum_{k=1}^{n}\frac{1}{2b_k}\left[2d_kA_k(x)(c_kA_k(x)-\xi_k)+ic_k(\partial_{x_k}A_k)(x)\right]\\
\omega_2(x) &= \sum_{k=1}^{n}\frac{1}{2b_k}\left[d_k^2A_k(x)^2+id_k(\partial_{x_k}A_k)(x)\right].
\end{aligned}
\tag{12.9}
$$

In particular, the conditions (2.6) and (11.1) are satisfied, and thus, we also have,

THEOREM 12.2. *Let $P(h)$ be as in (12.6) with V given by (12.2) and (12.4), $A_1,\dots,A_n\in C_b^\infty(\mathbb{R}^n;\mathbb{R})$, and $B_1,\dots,B_p\in C_b^\infty(\mathbb{R}^{n+p};\mathbb{R})$. Assume also (H3') and $L=1$. Then, the conclusions of Corollary 2.6 and Theorem 11.3 are valid for $P=P(h)$.*

Moreover, concerning the symbol of the effective Hamiltonian, in that case we have,

THEOREM 12.3. *Let $P(h)$ be as in (12.6) with V given by (12.2) and (12.4), $A_1,\dots,A_n\in C_b^\infty(\mathbb{R}^n;\mathbb{R})$, and $B_1,\dots,B_p\in C_b^\infty(\mathbb{R}^{n+p};\mathbb{R})$. Assume also (H3') and $L=1$. Then, the symbol $a(x,\xi;h)$ of the effective Hamiltonian verifies,*

$$
a(x,\xi;h)=a_0(x,\xi)+ha_1(x,\xi)+h^2a_2(x,\xi)+\mathcal{O}(h^3),
$$

with, for $(x,\xi)\in T^*(\Omega)$,

$$\begin{aligned}
a_0(x,\xi) &= \omega_0(x,\xi)+\lambda_{L'+1}(x)+W(x);\\
a_1(x,\xi) &= \omega_1(x,\xi)-i\nabla_\xi\omega_0(x,\xi)\langle\nabla_x u(x),u(x)\rangle\\
a_2(x,\xi) &= \sum_{k=1}^n\frac{1}{2b_k}\langle(\xi_k-d_kA_k(x))^2u(x),u(x)\rangle\\
&+\sum_{k,\ell=1}^n\frac{1}{b_kb_\ell}(\xi_k-c_kA_k)(\xi_\ell-c_\ell A_\ell)\langle R'(x,\lambda(x))\nabla_{x_k}u,\nabla_{x_\ell}u\rangle,
\end{aligned}$$

where ω_0 *and* ω_1 *are defined in (12.9), and*

$$R'(x,\lambda(x)):=\Pi_0^\perp(x)(\lambda(x)-Q(x))^{-1}\Pi_0^\perp(x),$$

is the reduced resolvent of $Q(x)$.

Proof – A possible proof may consist in using Proposition 10.2. Then, observing (with the notations of (12.8)) that, by definition,

$$\mathcal{J}=U_{x_0}D_xU_{x_0}^{-1}-D_x, \tag{12.10}$$

and, exploiting the fact that the $(L'+1)$-th normalized eigenstate $u(x)$ of $Q(x)$ is a twice differentiable function of x with values in $L^2(I\!R^{2p})$ (see , e.g., [**CoSe**], but this is also an easy consequence of (12.10) and the fact that $x\mapsto U_{x_0}(x)u(x)$ is smooth), and setting $v(x)=U_{x_0}(x)u(x)$, one can write,

$$\langle\mathcal{J}v,v\rangle_{\mathcal{H}}=\langle D_xu,u\rangle_{\mathcal{H}}-\langle D_xv,v\rangle_{\mathcal{H}}.$$

As a consequence, one also finds,

$$\sum_{k=1}^n\frac{1}{b_k}(\xi_k-c_kA_k)\langle\mathcal{J}_kv,v\rangle_{\mathcal{H}}-i\langle\nabla_\xi\omega_0\nabla_xv,v\rangle_{\mathcal{H}}=-i\langle\nabla_\xi\omega_0\nabla_xu,u\rangle_{\mathcal{H}}.$$

where ω_ℓ $(0\le\ell\le2)$ are defined in (12.9), and this permits to make appear many cancellations in the expression of $a(x,\xi;h)$ given in Proposition 10.2, leading to the required formulas.

However, there is a much simpler way to prove it, using directly the expressions (10.8) given in Proposition 10.2 for the operator A. Indeed, since in our case $x\mapsto u(x)$ is twice differentiable, for all $w\in L^2(I\!R^{n+p})$, we can write,

$$[D_x,\tilde\Pi_0]w=-i\langle w,\nabla_xu(x)\rangle u(x)-i\langle w,u(x)\rangle\nabla_xu(x),$$

and, for all $w\in C^1(I\!R^{3n};L^2(I\!R^{3p}))$,

$$\begin{aligned}
[D_x^2,\tilde\Pi_0]w &= [D_x,\tilde\Pi_0]\cdot D_xw+D_x\cdot[D_x,\tilde\Pi_0]w\\
&= -2i\langle D_xw,\nabla_xu(x)\rangle u(x)-2i\langle D_xw,u(x)\rangle\cdot\nabla_xu(x)\\
&\quad -\langle w,\nabla_xu(x)\rangle\cdot\nabla_xu(x)-\langle w,u(x)\rangle\nabla_x\cdot\nabla_xu(x).
\end{aligned}$$

This permits to write explicitly the operator $[\tilde{\Pi}_0, \tilde{P}] = [\tilde{\Pi}_0, \boldsymbol{\omega}]$ as,

$$\begin{aligned}
[\tilde{\Pi}_0, \tilde{P}]w &= ih\sum_{k=1}^{n}\frac{1}{b_k}\langle(hD_{x_k}-(c_k+hd_k)A_k)w, \nabla_{x_k}u(x)\rangle u(x)\\
&+ih\sum_{k=1}^{n}\frac{1}{b_k}\langle(hD_{x_k}-(c_k+hd_k)A_k)w, u(x)\rangle\cdot\nabla_{x_k}u(x)\\
&+h^2\sum_{k=1}^{n}\frac{1}{2b_k}\left(\langle w, \nabla_{x_k}u(x)\rangle\cdot\nabla_{x_k}u(x)+\langle w, u(x)\rangle\nabla_{x_k}^2u(x)\right).
\end{aligned}$$

In particular, taking $w = Z_1^*\alpha(x) = \alpha(x)u(x)$, $\alpha \in H^1(I\!R^{3n}))$, and using the fact that $R'(x, \lambda(x))u(x) = 0$, one finds,

$$\begin{aligned}
R'(x,\lambda(x))[\tilde{\Pi}_0, \tilde{P}]Z_1^*\alpha &= ih\sum_{k=1}^{n}\frac{1}{b_k}\left((hD_{x_k}-c_kA_k)\alpha\right)R'(x,\lambda(x))\nabla_{x_k}u(x)\\
&\qquad +\mathcal{O}(h^2\|\alpha\|),
\end{aligned}$$

and then,

$$\begin{aligned}
&Z_1[\tilde{P}, \tilde{\Pi}_0]R'(x,\lambda(x))[\tilde{\Pi}_0, \tilde{P}]Z_1^*\alpha\\
&= h^2\sum_{k,\ell=1}^{n}\frac{1}{b_kb_\ell}\left((hD_{x_k}-c_kA_k)(hD_{x_\ell}-c_\ell A_\ell)\alpha\right)\times\\
&\qquad\times\langle R'(x,\lambda(x))\nabla_{x_k}u(x), \nabla_{x_\ell}u(x)\rangle+\mathcal{O}(h^3\|\alpha\|),
\end{aligned}$$

This obviously permits to compute the principal symbol of the partial differential operator A_2 appearing in (10.8). The (full) symbol of $A_1 = Z_1\tilde{P}Z_1^*$ is even easier to compute, and the result follows. •

REMARK 12.4. *The smoothness with respect to x of all the coefficients appearing in $a(x,\xi;h)$ is a priori known, but can also be recovered directly by using (12.10). For instance, writing $\langle\nabla_x u(x), u(x)\rangle$ as,*

$$\langle\nabla_x u(x), u(x)\rangle = \langle\nabla_x U_{x_0}u(x), U_{x_0}u(x)\rangle + i\langle\mathcal{J}(x)U_{x_0}u(x), U_{x_0}u(x)\rangle,$$

permits to see its smoothness near x_0.

REMARK 12.5. *Using the expression of A_3 appearing in (10.8), one could also compute the next term (i.e., the h^3-term) in $a(x,\xi;h)$.*

REMARK 12.6. *Analogous formulas can be obtained in a very similar way in the case where L is arbitrary but $\lambda_{L'+1} = \cdots = \lambda_{L'+L}$.*

REMARK 12.7. *Although we did not do it here, we can also treat the case of unbounded magnetic potential (e.g., constant magnetic field). Then, the estimates on the coefficients c_α's in Assumption (H1) are not satisfied anymore, but, since we mainly work in a compact region of the x-space, it is clear that an adaptation of our arguments lead to the same results.*

REMARK 12.8. *In the case of a free molecule (or, more generally, if the external electromagnetic field is invariant under the translations of the type $(x, y) \mapsto (x_1 + \alpha, \ldots, x_n + \alpha, y_1 + \alpha, \ldots, y_p + \alpha)$ for any $\alpha \in I\!R^3$), one can factorize the quantum motion, e.g., by using the so-called center of mass of the nuclei coordinate system,* *as in* [**KMSW**]. *Then, denoting by R the position of the center of mass of the nuclei, the operator takes the form,*

$$P(h) = H_0(D_R) + P'(h) + h^2 p(D_y),$$

where $H_0(D_R)$ stands for the quantum-kinetic energy of the center of mass of the nuclei, $P'(h)$ has a form similar to that of $P(h)$ in (12.6) (but now, with $x \in I\!R^{3(n-1)}$ denoting the relative *positions of the nuclei), and $p(D_y)$ is a PDO of order 2 with respect to y, with constant coefficients (the so-called* isotopic term*). Therefore, one obtains the factorization,*

$$e^{-itP(h)/h} = e^{-itH_0(D_R)/h} e^{-it(P'(h)+h^2 p(D_y))/h}, \tag{12.11}$$

and it is easy to verify that our previous constructions can be performed with $Q(x)$ replaced by $Q(x) + h^2 p(D_y)$. In particular, under the same assumptions as in Theorem 12.1, the quantum evolution under $P'(h) + h^2 p(D_y)$ of an initial state φ_0 verifying (2.4) with P replaced by $P'(h)$ (that is, a much weaker assumption) can be expressed in terms of the quantum evolution associated to a $L \times L$ matrix of h-admissible operators on $L^2(I\!R^{3(n-1)})$. In that case, (12.11) provides a way to reduce the evolution of φ_0 under $P(h)$, too.

APPENDIX A

Smooth Pseudodifferential Calculus with Operator-Valued Symbol

We recall the usual definition of h-admissible operator with operator-valued symbol. In some sense, this corresponds to a simple case of the more general definitions given in [**Ba, GMS**]. For $m \in \mathbb{R}$ and $\mathcal{H}$ a Hilbert space, we denote by $H^m(\mathbb{R}^n;\mathcal{H})$ the standard m-th order Sobolev space on $\mathbb{R}^n$ with values in $\mathcal{H}$.

DEFINITION A.1. *Let $m \in \mathbb{R}$ and let $\mathcal{H}_1$ and $\mathcal{H}_2$ be two Hilbert space. An operator $A = A(h) : H^m(\mathbb{R}^n;\mathcal{H}_1) \to L^2(\mathbb{R}^n;\mathcal{H}_2)$ with $h \in (0,h_0]$ is called h-admissible (of degree m) if, for any $N \geq 1$,*

$$A(h) = \sum_{j=0}^{N} h^j \operatorname{Op}_h(a_j(x,\xi;h)) + h^N R_N(h), \tag{A.1}$$

where R_N is uniformly bounded from $H^m(\mathbb{R}^n;\mathcal{H}_1)$ to $L^2(\mathbb{R}^n;\mathcal{H}_2)$ for $h \in (0,h_0]$, and, for all $h > 0$ small enough, $a_j \in C^\infty(T^*\mathbb{R}^n;\mathcal{L}(\mathcal{H}_1;\mathcal{H}_2))$, with

$$\|\partial^\alpha a_j(x,\xi;h)\|_{\mathcal{L}(\mathcal{H}_1;\mathcal{H}_2)} \leq C_\alpha \langle\xi\rangle^m \tag{A.2}$$

for all $\alpha \in {\mathbb{Z}_+}^{2n}$ and some positive constant C_α, uniformly for $(x,\xi) \in T^*\mathbb{R}^n$ and $h > 0$ small enough. In that case, the formal series,

$$a(x,\xi;h) = \sum_{j\geq 0} h^j a_j(x,\xi;h), \tag{A.3}$$

is called the symbol of A (it can be resummed up to a remainder in $\mathcal{O}(h^\infty\langle\xi\rangle^m)$ together with all its derivatives). Moreover, in the case $m = 0$ and $\mathcal{H}_2 = \mathcal{H}_1$, A is called a (bounded) h-admissible operator on $L^2(\mathbb{R}^n;\mathcal{H}_1)$.

Here, we have denoted by $\operatorname{Op}_h(a)$ the standard quantization of a symbol a, defined by the following formula:

$$\operatorname{Op}_h(a)u(x) := \frac{1}{(2\pi h)^n}\int e^{i(x-y)\xi/h} a(x,\xi)\, u(y)dyd\xi, \tag{A.4}$$

valid for any tempered distribution u, and where the integral has to be interpreted as an oscillatory one. Actually, by the Calderón-Vaillancourt Theorem (see, e.g., [**GMS, DiSj1, Ma2, Ro1**], and below), the estimate (A.2) together with the quantization formula (A.4), permit to define $\operatorname{Op}_h(a)$ as a bounded operator $H^m(\mathbb{R}^n;\mathcal{H}_1) \to L^2(\mathbb{R}^n;\mathcal{H}_2)$. Let us also observe that, very often, the formal series (A.3) are indeed identified with one of their resummations (and thus, the symbol is considered as a function, rather than a formal series). Indeed, since the various resummations (together with all their derivatives) differ by uniformly $\mathcal{O}(h^\infty\langle\xi\rangle^m)$ terms, in view of (A.1) and the Calderón-Vaillancourt Theorem, it is clear that this has no real importance.

As it is well known (see, e.g., [**Ba, DiSj1, GMS, Ma2**]), with such a type of quantization is associated a full and explicit symbolic calculus that permits to handle these operators in a very easy and pleasant way. In particular, we have the following results:

PROPOSITION A.2 (Composition). *Let A and B be two bounded h-admissible operators on $L^2(\mathbb{R}^n;\mathcal{H}_1)$, with respective symbols a and b. Then, the composition $A\circ B$ is an h-admissible operators on $L^2(\mathbb{R}^n;\mathcal{H}_1)$, too, and its symbol $a\sharp b$ is given by the formal series,*

$$a\sharp b(x,\xi;h)=\sum_{\alpha\in\mathbb{Z}_+^n}\frac{h^{|\alpha|}}{i^{|\alpha|}\alpha!}\partial_\xi^\alpha a(x,\xi;h)\partial_x^\alpha b(x,\xi;h).$$

REMARK A.3. *There is a similar result for the composition of unbounded h-admissible operators, but it requires more conditions on the remainder $R_N(h)$ appearing in (A.1) (see [**Ba, GMS**]).*

PROPOSITION A.4 (Parametrix). *Let A be a bounded h-admissible operator on $L^2(\mathbb{R}^n;\mathcal{H}_1)$, such that any resummation a of its symbol is elliptic, in the sense that $a(x,\xi;h)$ is invertible on $\mathcal{H}_1$ for any $(x,\xi;h)$, and its inverse verifies,*

$$\|a(x,\xi;h)^{-1}\|_{\mathcal{L}(\mathcal{H}_1)}=\mathcal{O}(1),$$

uniformly for $(x,\xi)\in T^*\mathbb{R}^n$ and $h>0$ small enough. Then, A is invertible on $L^2(\mathbb{R}^n;\mathcal{H}_1)$, its inverse A^{-1} is h-admissible, and its symbol b verifies,

$$b=a^{-1}+hr,$$

with $r=\sum_{j\geq 0}h^jr_j$, $\|\partial^\alpha r_j\|_{\mathcal{L}(\mathcal{H}_1)}=\mathcal{O}(1)$ uniformly.

REMARK A.5. *It is easy to see that the ellipticity of any resummation of the symbol is equivalent to the ellipticity of the function $a_0(x,\xi;h)$ appearing in (A.1) (and thus, to the ellipticity of at least one resummation).*

REMARK A.6. *Of course, the r_j's can actually be all determined recursively, by using the identity $a\sharp b=1$ (this gives a possible choice for them, but this choice is not unique since we have allowed them to depend on h).*

PROPOSITION A.7 (Functional Calculus). *Let A be a self-adjoint h-admissible operator on $L^2(\mathbb{R}^n;\mathcal{H}_1)$, and let $f\in C_0^\infty(\mathbb{R})$. Then, $f(A)$ is h-admissible, and its symbol b verifies,*

$$b=f(\operatorname{Re}a)+hr,$$

where $\operatorname{Re}a:=(a+a^*)/2$, and $r=\sum_{j\geq 0}h^jr_j$, $\|\partial^\alpha r_j\|_{\mathcal{L}(\mathcal{H}_1)}=\mathcal{O}(1)$ uniformly.

PROPOSITION A.8 (Calderón-Vaillancourt Theorem). *Let $a=a(x,\xi)$ be in $C^\infty(T^*\mathbb{R}^n;\mathcal{L}(\mathcal{H}_1;\mathcal{H}_2))$, such that, for all $\alpha\in\mathbb{Z}_+^{2n}$, $\|\partial^\alpha a(x,\xi)\|_{\mathcal{L}(\mathcal{H}_1;\mathcal{H}_2)}$ is uniformly bounded on $T^*\mathbb{R}^n$. Then, $\mathrm{Op}_h(a)$ (defined, e.g., on $\mathcal{S}(\mathbb{R}^n;\mathcal{H}_1)$) extends to a bounded operator : $L^2(\mathbb{R}^n;\mathcal{H}_1)\to L^2(\mathbb{R}^n;\mathcal{H}_2)$, and there exist two constants C_n and M_n, depending only on the dimension n, such that,*

$$\|\mathrm{Op}_h(a)\|_{\mathcal{L}(L^2(\mathbb{R}^n;\mathcal{H}_1);L^2(\mathbb{R}^n;\mathcal{H}_2))}\leq C_n\sum_{|\alpha|\leq M_n}\sup_{T^*\mathbb{R}^n}|\partial^\alpha a(x,\xi)|.$$

APPENDIX B

Propagation of the Support

THEOREM B.1. *Let P be as in (2.2) with (H1)-(H2), and let K_0 be a compact subset of $\mathbb{R}^n_x$, $f \in C_0^\infty(\mathbb{R})$ and $\varphi_0 \in L^2(\mathbb{R}^n;\mathcal{H})$, such that $\|\varphi_0\| = 1$, and,*

$$\|(1-f(P))\varphi_0\|_{L^2(\mathbb{R}^n;\mathcal{H})} + \|\varphi_0\|_{L^2(K_0^c;\mathcal{H})} = \mathcal{O}(h^\infty).$$

Then, for any $\varepsilon > 0$, any $T > 0$, and any $g \in C_0^\infty(\mathbb{R})$ such that $gf = f$, the compact set defined by,

$$K_{T,\varepsilon} := \{x \in \mathbb{R}^n\ ;\ \text{dist}\,(x, K_0) \le \varepsilon + C_1 T\},$$

with

$$C_1 := \frac{1}{2}\|\nabla_\xi \omega(x, hD_x) g(P)\|,$$

verifies,

$$\sup_{t\in[0,T]} \|e^{-itP/h}\varphi_0\|_{L^2(K_{T,\varepsilon}^c;\mathcal{H})} = \mathcal{O}(h^\infty),$$

as $h \to 0$.

Proof – First, we need the following lemma:

LEMMA B.2. *For any $\chi \in C_b^\infty(\mathbb{R}^n)$, such that $\text{supp}\chi \subset K_0^c$, and for any $g \in C_0^\infty(\mathbb{R})$, one has,*

$$\|\chi(x)g(P)\varphi_0\| = \mathcal{O}(h^\infty).$$

Proof – Consider a sequence $(\chi_j)_{j\in\mathbb{N}} \subset C_b^\infty(\mathbb{R}^n)$, $\text{supp}\chi_j \subset K_0^c$ and such that

$$\chi_{j+1}\chi_j = \chi_j, \quad \chi_j\chi = \chi.$$

Then, in view of (4.8), it is sufficient to show that, for any $N \ge 0$,

$$\|\chi_j(x)(P-\lambda)^{-1}\varphi_0\| = \mathcal{O}(h^N |\operatorname{Im}\lambda|^{-(N+1)}),$$

uniformly as $h, |\operatorname{Im}\lambda| \to 0_+$.

We set, $u_j = \chi_j(x)(P-\lambda)^{-1}\varphi_0$, and we observe that, for all $j \in \mathbb{N}$, one has $\|u_j\| = \mathcal{O}(|\operatorname{Im}\lambda|^{-1})$. By induction on N, let us suppose, for all $j \in \mathbb{N}$,

$$\|\chi_j(x)(P-\lambda)^{-1}\varphi_0\| = \mathcal{O}(h^N |\operatorname{Im}\lambda|^{-(N+1)}).$$

Since $\chi_{j+1} = 1$ on $\text{Supp}\,\chi_j$, and P is differential in x, we have,

$$(P-\lambda)u_j = \chi_j\varphi_0 + [P,\chi_j]\chi_{j+1}(P-\lambda)^{-1}\varphi_0,$$

and thus,

$$u_j = (P-\lambda)^{-1}\chi_j\varphi_0 + (P-\lambda)^{-1}[\boldsymbol{\omega},\chi_j]u_{j+1}.$$

Now, by assumption, we have $\|\chi_j\varphi_0\| = \mathcal{O}(h^\infty)$, and therefore, $\|(P-\lambda)^{-1}\chi_j\varphi_0\| = \mathcal{O}(h^\infty|\operatorname{Im}\lambda|^{-1})$. Moreover, using (H1)-(H2), it is easy to see that the operator

$|\operatorname{Im}\lambda|h^{-1}(P-\lambda)^{-1}[\boldsymbol{\omega},\chi_j]$ is uniformly bounded on $L^2(I\!R^n;\mathcal{H})$. Hence, using the induction hypothesis, we obtain,

$$\|u_j\| = \mathcal{O}(h^{\infty}|\operatorname{Im}\lambda|^{-1}) + \mathcal{O}(h^{N+1}|\operatorname{Im}\lambda|^{-(N+2)}) = \mathcal{O}(h^{N+1}|\operatorname{Im}\lambda|^{-(N+2)})$$

for any $j\in I\!N$, and the lemma follows. •

Now, for any $F\in C^{\infty}(I\!R_+\times I\!R^n_x;I\!R)$, let us compute the quantity,

$$\begin{aligned}
&\partial_t\langle F(t,x)f(P)e^{-itP/h}\varphi_0, f(P)e^{-itP/h}\varphi_0\rangle\\
&\quad= \operatorname{Re}\langle(\partial_t F - ih^{-1}FP)f(P)e^{-itP/h}\varphi_0, f(P)e^{-itP/h}\varphi_0\rangle\\
&\quad= \langle(\partial_t F - \frac{i}{2h}[F,P])f(P)e^{-itP/h}\varphi_0, f(P)e^{-itP/h}\varphi_0\rangle\\
&\quad= \langle(\partial_t F + \frac{i}{2h}[\boldsymbol{\omega},F])f(P)e^{-itP/h}\varphi_0, f(P)e^{-itP/h}\varphi_0\rangle. \qquad \text{(B.1)}
\end{aligned}$$

Then, we fix $g\in C_0^{\infty}(I\!R)$ such that $gf=f$, and, for $j\in I\!N$, we set,

$$F_j(t,x) := \varphi_j(\operatorname{dist}(x,K_0) - C_1 t), \qquad \text{(B.2)}$$

where $C_1 = \frac{1}{2}\|\nabla_\xi\omega(x,hD_x)g(P)\|$, and the φ_j's are in $C_b^{\infty}(I\!R;I\!R_+)$ with support in $[\varepsilon,+\infty)$, verify $\varphi_j(s)=1$ for $s\geq\varepsilon+\frac{1}{j}$, $\varphi_{j+1}=1$ near $\operatorname{Supp}\varphi_j$, and are such that,

$$\varphi_j' := \phi_j^2 \geq 0 \text{ with } \phi_j\in C_b^{\infty}(I\!R;I\!R).$$

In particular, $F_j\in C_b^{\infty}(I\!R_+\times I\!R^n_x;I\!R_+)$, and, setting $d(x):=\operatorname{dist}(x,K_0)$, we have,

$$\nabla_x F_j = \varphi_j'(d(x)-C_1t)\nabla d(x), \quad \partial_t F_j = -C_1\varphi_j'(d(x)-C_1t).$$

Moreover, since $\boldsymbol{\omega}=\omega(x,hD_x)$ is a differential operator with respect x, of degree m, we see that,

$$\frac{i}{h}[\boldsymbol{\omega},F_j] = \nabla_x F_j\cdot\nabla_\xi\omega(x,hD_x) + hR_j, \qquad \text{(B.3)}$$

where $R_j = R_j(t,x,hD_x)$ is a differential operator of degree $m-2$ in x, with coefficients in $C_b^{\infty}(I\!R_+\times I\!R^n_x)$ and supported in $\{F_{j+1}=1\}$.

LEMMA B.3. *For any $N\geq 1$,*

$$\|R_jf(P)u\| = \mathcal{O}(\sum_{k=0}^{N} h^k\|F_{j+k+1}f(P)u\| + h^{N+1}\|u\|).$$

Proof – We write,

$$R_jf(P) = R_jF_{j+1}f(P) = R_jg(P)F_{j+1}f(P) + R_j[F_{j+1},g(P)]f(P).$$

Then, using (4.8) and the fact that $[P,F_{j+1}]=[\boldsymbol{\omega},F_{j+1}]$, we obtain,

$$\begin{aligned}
&R_j[F_{j+1},g(P)]\\
&\quad= \frac{1}{\pi}\int\bar\partial\tilde g(z)R_j(P-z)^{-1}[\boldsymbol{\omega},F_{j+1}](P-z)^{-1}dz\,d\bar z\\
&\quad= \frac{1}{\pi}\int\bar\partial\tilde g(z)R_j(P-z)^{-1}[\boldsymbol{\omega},F_{j+1}]F_{j+2}(P-z)^{-1}dz\,d\bar z\\
&\quad= \frac{1}{\pi}\int\bar\partial\tilde g(z)R_j(P-z)^{-1}[\boldsymbol{\omega},F_{j+1}](P-z)^{-1}F_{j+2}dz\,d\bar z\\
&\qquad+\frac{1}{\pi}\int\bar\partial\tilde g(z)R_j(P-z)^{-1}[\boldsymbol{\omega},F_{j+1}](P-z)^{-1}[\boldsymbol{\omega},F_{j+2}](P-z)^{-1}dz\,d\bar z,
\end{aligned}$$

and thus, by iteration,

$$\begin{aligned}
&R_j[F_{j+1}, g(P)]\\
&\quad = \sum_{k=1}^{N} \frac{1}{\pi} \int \overline{\partial}\tilde{g}(z) R_j (P-z)^{-1} \left(\prod_{\ell=1}^{k} \left([\boldsymbol{\omega}, F_{j+\ell}](P-z)^{-1} \right) \right) F_{j+k+1} dz\, d\bar{z}\\
&\qquad + \frac{1}{\pi} \int \overline{\partial}\tilde{g}(z) R_j (P-z)^{-1} \prod_{\ell=1}^{N+1} \left([\boldsymbol{\omega}, F_{j+\ell}](P-z)^{-1} \right) dz\, d\bar{z}.
\end{aligned}$$

Since $\|R_j(P-z)^{-1}\| = \mathcal{O}(1)$ and $\|[\boldsymbol{\omega}, F_{j+\ell}](P-z)^{-1}\| = \mathcal{O}(h)$, the result follows. •

As a consequence, we deduce from (B.3),

$$\begin{aligned}
&\frac{i}{h}[\boldsymbol{\omega}, F_j] f(P) e^{-itP/h} \varphi_0\\
&= \varphi_j'(d(x) - C_1 t) \nabla d(x) \nabla_\xi \omega(x, hD_x) f(P) e^{-itP/h} \varphi_0\\
&\qquad\qquad + \mathcal{O}\Big(\sum_{k=0}^{N} h^{k+1} \|F_{j+k+1} f(P) e^{-itP/h} \varphi_0\| + h^{N+2}\Big)\\
&= \phi_j(d(x) - C_1 t) \nabla d(x) \nabla_\xi \omega(x, hD_x) g(P) \phi_j(d(x) - C_1 t) f(P) e^{-itP/h} \varphi_0\\
&\quad + \phi_j(d(x) - C_1 t)) \nabla d(x) [\phi_j(d(x) - C_1 t), \nabla_\xi \omega(x, hD_x)] f(P) e^{-itP/h} \varphi_0\\
&\quad + \phi_j(d(x) - C_1 t)) \nabla d(x) \nabla_\xi \omega(x, hD_x) [\phi_j(d(x) - C_1 t), g(P)] f(P) e^{-itP/h} \varphi_0\\
&\qquad\qquad + \mathcal{O}\Big(\sum_{k=0}^{N} h^{k+1} \|F_{j+k+1} f(P) e^{-itP/h} \varphi_0\| + h^{N+2}\Big),
\end{aligned}$$

and thus, since ϕ_j is supported in $\{F_{j+1} = 1\}$, as in the proof of Lemma B.3, we obtain,

$$\begin{aligned}
&\frac{i}{h}[\boldsymbol{\omega}, F_j] f(P) e^{-itP/h} \varphi_0\\
&\quad = \phi_j(d(x) - C_1 t) \nabla d(x) \nabla_\xi \omega(x, hD_x) g(P) \phi_j(d(x) - C_1 t) f(P) e^{-itP/h} \varphi_0\\
&\qquad\qquad + \mathcal{O}\Big(\sum_{k=0}^{N} h^{k+1} \|F_{j+k+1} f(P) e^{-itP/h} \varphi_0\| + h^{N+2}\Big),
\end{aligned}$$

for any fixed $N \geq 1$.

Going back to (B.1), and using the fact that $\|\nabla d(x) \nabla_\xi \omega(x, hD_x) g(P)\| \leq C_1$, this gives,

$$\begin{aligned}
&\partial_t \langle F_j(t,x) f(P) e^{-itP/h} \varphi_0, f(P) e^{-itP/h} \varphi_0 \rangle\\
&\qquad \leq \mathcal{O}\Big(\sum_{k=0}^{N} h^{k+1} \|F_{j+k+1} f(P) e^{-itP/h} \varphi_0\|^2 + h^{N+2}\Big),
\end{aligned}$$

and therefore, integrating between 0 and t, and using Lemma B.2,

$$\begin{aligned}
&\langle F_j(t,x) f(P) e^{-itP/h} \varphi_0, f(P) e^{-itP/h} \varphi_0 \rangle\\
&\qquad = \mathcal{O}\Big(\sum_{k=0}^{N} h^{k+1} \int_0^t \|F_{j+k+1} f(P) e^{-isP/h} \varphi_0\|^2 ds + t h^{N+2}\Big),
\end{aligned}$$

In particular, since

$$\|F_j(t,x)f(P)e^{-itP/h}\varphi_0\|^2 \leq \langle F_j(t,x)f(P)e^{-itP/h}\varphi_0, f(P)e^{-itP/h}\varphi_0\rangle,$$

we have $\|F_j(t,x)f(P)e^{-itP/h}\varphi_0\|^2 = \mathcal{O}(h)$ for any $j \in \mathbb{N}$, and then, by induction, $\|F_j(t,x)f(P)e^{-itP/h}\varphi_0\|^2 = \mathcal{O}(h^N)$ for all $N \in \mathbb{N}$. Due to the definition (B.2) of F_j, this proves the theorem. •

APPENDIX C

Two Technical Lemmas

LEMMA C.1. *Let $\psi_j, \chi_j \in C_0^\infty(\mathbb{R}^n)$, such that $\chi_j = 1$ near $\operatorname{Supp}\psi_j$. Then, for any $f \in C_0^\infty(\mathbb{R})$, one has,*

$$U_j\psi_j f(\tilde{P})U_j^{-1}\chi_j = \psi_j f(U_j\chi_j\tilde{P}U_j^{-1}\chi_j) + \mathcal{O}(h^\infty).$$

Proof – By (4.8), and taking the adjoint, it is enough to prove, for any $N \geq 1$,

$$U_j\chi_j(\tilde{P}-z)^{-1}U_j^{-1}\psi_j = (U_j\chi_j\tilde{P}U_j^{-1}\chi_j - z)^{-1}\psi_j + \mathcal{O}(h^N|\operatorname{Im} z|^{-N'}),$$

locally uniformly for $z \in \mathbb{C}$, and with some $N' = N'(N) < +\infty$. Let $v \in L^2(\mathbb{R}^n)$ and set $u := (\tilde{P}-z)^{-1}U_j^{-1}\psi_j v$. By Lemma 4.11 (and its proof), we know that,

$$u = \chi_j u + \mathcal{O}(h^N|\operatorname{Im} z|^{-N'}\|v\|), \tag{C.1}$$

for some $N' = N'(N) < +\infty$. On the other hand, we have,

$$\begin{aligned}(U_j\chi_j\tilde{P}U_j^{-1}\chi_j - z)U_j\chi_j u &= U_j\chi_j\tilde{P}u - zU_j\chi_j u + U_j\chi_j\tilde{P}(\chi_j^2-1)u\\ &= U_j\chi_j(zu + U_j^{-1}\psi_j v) - zU_j\chi_j u + U_j\chi_j\tilde{P}(\chi_j^2-1)u\\ &= \psi_j v + U_j\chi_j\tilde{P}(\chi_j^2-1)u,\end{aligned}$$

and thus, using (C.1),

$$\begin{aligned}U_j\chi_j u &= (U_j\chi_j\tilde{P}U_j^{-1}\chi_j - z)^{-1}(\psi_j v + U_j\chi_j\tilde{P}(\chi_j^2-1)u)\\ &= (U_j\chi_j\tilde{P}U_j^{-1}\chi_j - z)^{-1}\psi_j v + \mathcal{O}(h^N|\operatorname{Im} z|^{-N''}\|v\|),\end{aligned}$$

for some other $N'' = N''(N) < +\infty$. Then, the result follows. •

LEMMA C.2. *Let $\psi, \chi \in C_0^\infty(\mathbb{R}^n)$, such that $\chi = 1$ near $\operatorname{Supp}\psi$. Then, for any $\rho \in C_0^\infty(\mathbb{R})$, one has,*

$$\rho(\chi\omega\chi)\psi = \rho(\omega)\psi + \mathcal{O}(h^\infty).$$

Proof – The proof is very similar to (but simpler than) the one of Lemma C.1, and we omit it. •

Bibliography

[Ba] A. Balazard-Konlein. *Calcul fonctionel pour des opérateurs h-admissibles à symbole opérateurs et applications*, PhD Thesis, Université de Nantes (1985).

[Be] M. V. Berry, *The Quantum Phase, Five Years After*, in Geometric Phases in Physics (A. Shapere and F. Wilczek, Eds), World Scientific, Singapore, 1989.

[BoOp] M. Born, R. Oppenheimer, *Zur Quantentheorie der Molekeln*, Ann. Phys. **84** (1927), 457–484.

[BrNo] R. Brummelhuis, J. Nourrigat, *Scattering amplitude for Dirac operators*, Comm. Partial Differential Equations **24**(1-2) (1999),377–394.

[CDS] J.-M. Combes, P. Duclos, R. Seiler, *The Born-Oppenheimer approximation*, in: Rigorous atomic and molecular physics, G. Velo and A. Wightman (Eds.), Plenum Press New-York (1981), 185-212.

[CoRo] M. Combescure, D. Robert, *Semiclassical spreading of quantum wave packets and applications near unstable fixed points of the classical flow*, Asymptotic Analysis **14** (1997), 377–404.

[CoSe] J.-M. Combes, R. Seiler, *Regularity and asymptotic properties of the discrete spectrum of electronic Hamiltonians*, Int. J. Quant. Chem. **XIV** (1978), 213-229.

[DiSj1] M.Dimassi, J.Sjöstrand, *Spectral asymptotics in the semi-classical limit.* London Mathematical Society Lecture Note Series, **268**, Cambridge University Press, Cambridge, (1999).

[GMS] C. Gérard, A. Martinez, J. Sjöstrand, *A mathematical approach to the effective hamiltonian in perturbed periodic problems*, Comm. Math. Physics **142** (2),(1991), 217–244.

[GuSt] V. Guillemin, S. Sternberg, *Geometric Asymptotics*, Amer. Math. Soc. Survey **14**, (1977)

[Ha1] G. Hagedorn, *Semiclassical quantum mechanics IV*, Ann. Inst. H. Poincaré **42** (1985), 363–374.

[Ha2] G. Hagedorn, *High order corrections to the time-independent Born-Oppenheimer approximation I: Smooth potentials*, Ann. Inst. H. Poincaré **47** (1987), 1–16.

[Ha3] G. Hagedorn, *High order corrections to the time-independent Born-Oppenheimer approximation II: Diatomic Coulomb systems*, Comm. Math. Phys. **116**, (1988), 23–44.

[Ha4] G. Hagedorn, *A time-dependent Born-Oppenheimer approximation*, Comm. Math. Phys. **77**, (1980), 1–19.

[Ha5] G. Hagedorn, *High order corrections to the time-dependent Born-Oppenheimer approximation I: Smooth potentials*, Ann. Math. **124**, 571–590 (1986). Erratum. Ann. Math, **126**, 219 (1987)

[Ha6] G. Hagedorn, *High order corrections to the time-dependent Born-Oppenheimer approximation II: Coulomb systtems*, Comm. Math. Phys. **116**, (1988), 23–44.

[HaJo] G. Hagedorn, A. Joye, *A Time-Dependent Born–Oppenheimer Approximation with Exponentially Small Error Estimates*, Comm. Math. Phys. **223** (3) (2001), 583–626.

[HeSj11] B. Helffer, J. Sjöstrand, *Multiple wells in the semiclassical limit I*, Comm. Part. Diff. Eq. **9** (4) (1984), 337–408.

[HeSj12] B. Helffer, J. Sjöstrand, *Semiclassical Analysis of Harper's Equation III*, Mem. Soc. Math. Fr., Nouv. Ser. **39**, (1989).

[Ka] T. Kato, *Perturbation Theory for Linear Operators*, 2nd ed. Classics in Mathematics, Springer-Verlag, Berlin (1980)

[KMSW] M. Klein, A. Martinez, R. Seiler, X.P. Wang, *On the Born-Oppenheimer Expansion for Polyatomic Molecules*, Commun. Math. Phys. **143**(3) (1992), 607–639

[Ma1] A. Martinez, *Développement asymptotiques et efffet tunnel dans l'approximation de Born-Oppenheimer*, Ann. Inst. H. Poincaré **49** (1989), 239–257.

[Ma2] A. Martinez, *An Introduction to Semiclassical and Microlocal Analysis*, Universitext. Springer-Verlag, New York, 2002.

[MaSo] A. Martinez, V. Sordoni, *A general reduction scheme for the time-dependent Born-Oppenheimer approximation*, C.R. Acad. Sci. Paris, Ser. I **334** (2002).

[MeSj1] A. Melin, J. Sjöstrand, *Fourier integral operators with complex phase functions and parametrix for an interior boundary value problem*, Commun. Partial Differ. Equations **1** (1976), 313–400

[Ne1] G. Nenciu, *Linear Adiabatic Theory, Exponential Estimates*, Commun. Math. Phys. **152** (1993), 479–496.

[Ne2] G. Nenciu, *On asymptotic perturbation theory for quantum mechanics: almost invariant subspaces and gauge invariant magnetic perturbation theory* J. Math. Phys. , **43** (2002), 1273-1298.

[NeSo] G. Nenciu, V. Sordoni, *Semiclassical limit for multistate Klein-Gordon systems: almost invariant subspaces and scattering theory*, J. Math. Phys. **45** (2004), 3676–3696

[PST] G. Panati, H. Spohn, S. Teufel, *Space-adiabatic perturbation theory*, Adv. Theor. Math. Phys. **7** (2003), 145–204.

[Ra] A. Raphaelian, *Ion-atom scattering within a Born-Oppenheimer framework*, Dissertation Technische Universität Berlin (1986)

[Ro1] D. Robert, *Autour de l'Approximation Semi-Classique*, Birkhäuser (1987)

[Ro2] D. Robert, *Remarks on asymptotic solutions for time dependent Schrödinger equations*, Optimal Control and Partial Differential Equations, IOS Press (2001)

[Sj1] J. Sjöstrand, *Singularités Analytiques Microlocales*, Astérisque **95** (1982)

[Sj2] J. Sjöstrand, *Projecteurs adiabatique du point de vue pseudodifferéntiel*, C. R. Acad. Sci. Paris **317**, Série I (1993), 217–220.

[So] V. Sordoni, *Reduction scheme for semiclassical operator-valued Schrödinger type equation and application to scattering*, Comm. Partial Differential Equations **28** (7-8) (2003), 1221–1236

[SpTe] H. Spohn, S. Teufel, *Adiabatic decoupling and time-dependent Born-Oppenheimer theory*, Comm. Math. Phys. **224** (1) (2001), 113–132.

[Te] S. Teufel, *Adiabatic Perturbation Theory in Quantum Dynamics*, Lecture Notes in Math. **1821**, Springer-Verlag, Berlin, Heidelberg, New York (2003)

Editorial Information

To be published in the *Memoirs*, a paper must be correct, new, nontrivial, and significant. Further, it must be well written and of interest to a substantial number of mathematicians. Piecemeal results, such as an inconclusive step toward an unproved major theorem or a minor variation on a known result, are in general not acceptable for publication.

Papers appearing in *Memoirs* are generally at least 80 and not more than 200 published pages in length. Papers less than 80 or more than 200 published pages require the approval of the Managing Editor of the Transactions/Memoirs Editorial Board.

As of March 31, 2009, the backlog for this journal was approximately 12 volumes. This estimate is the result of dividing the number of manuscripts for this journal in the Providence office that have not yet gone to the printer on the above date by the average number of monographs per volume over the previous twelve months, reduced by the number of volumes published in four months (the time necessary for preparing a volume for the printer). (There are 6 volumes per year, each usually containing at least 4 numbers.)

A Consent to Publish and Copyright Agreement is required before a paper will be published in the *Memoirs*. After a paper is accepted for publication, the Providence office will send a Consent to Publish and Copyright Agreement to all authors of the paper. By submitting a paper to the *Memoirs*, authors certify that the results have not been submitted to nor are they under consideration for publication by another journal, conference proceedings, or similar publication.

Information for Authors

Memoirs are printed from camera copy fully prepared by the author. This means that the finished book will look exactly like the copy submitted.

Initial submission. The AMS uses Centralized Manuscript Processing for initial submissions. Authors should submit a PDF file using the Initial Manuscript Submission form found at `www.ams.org/peer-review-submission`, or send one copy of the manuscript to the following address: Centralized Manuscript Processing, MEMOIRS OF THE AMS, 201 Charles Street, Providence, RI 02904-2294 USA. If a paper copy is being forwarded to the AMS, indicate that it is for it Memoirs and include the name of the corresponding author, contact information such as email address or mailing address, and the name of an appropriate Editor to review the paper (see the list of Editors below).

The paper must contain a *descriptive title* and an *abstract* that summarizes the article in language suitable for workers in the general field (algebra, analysis, etc.). The *descriptive title* should be short, but informative; useless or vague phrases such as "some remarks about" or "concerning" should be avoided. The *abstract* should be at least one complete sentence, and at most 300 words. Included with the footnotes to the paper should be the 2000 *Mathematics Subject Classification* representing the primary and secondary subjects of the article. The classifications are accessible from `www.ams.org/msc/`. The list of classifications is also available in print starting with the 1999 annual index of *Mathematical Reviews*. The Mathematics Subject Classification footnote may be followed by a list of *key words and phrases* describing the subject matter of the article and taken from it. Journal abbreviations used in bibliographies are listed in the latest *Mathematical Reviews* annual index. The series abbreviations are also accessible from `www.ams.org/msnhtml/serials.pdf`. To help in preparing and verifying references, the AMS offers MR Lookup, a Reference Tool for Linking, at `www.ams.org/mrlookup/`.

Electronically prepared manuscripts. The AMS encourages electronically prepared manuscripts, with a strong preference for AMS-LaTeX. To this end, the Society has prepared AMS-LaTeX author packages for each AMS publication. Author packages include instructions for preparing electronic manuscripts, samples, and a style file that generates the particular design specifications of that publication series. Though $\mathcal{AMS}$-LaTeX is the highly preferred format of TeX, author packages are also available in $\mathcal{AMS}$-TeX.

Authors may retrieve an author package for *Memoirs of the AMS* from `www.ams.org/journals/memo/memoauthorpac.html` or via FTP to `ftp.ams.org` (login as `anonymous`, enter username as password, and type `cd pub/author-info`). The *AMS Author Handbook* and the *Instruction Manual* are available in PDF format from the author package link. The author package can also be obtained free of charge by sending email to `tech-support@ams.org` (Internet) or from the Publication Division, American Mathematical Society, 201 Charles St., Providence, RI 02904-2294, USA. When requesting an author package, please specify $\mathcal{AMS}$-LaTeX or $\mathcal{AMS}$-TeX and the publication in which your paper will appear. Please be sure to include your complete mailing address.

After acceptance. The final version of the electronic file should be sent to the Providence office (this includes any TeX source file, any graphics files, and the DVI or PostScript file) immediately after the paper has been accepted for publication.

Before sending the source file, be sure you have proofread your paper carefully. The files you send must be the EXACT files used to generate the proof copy that was accepted for publication. For all publications, authors are required to send a printed copy of their paper, which exactly matches the copy approved for publication, along with any graphics that will appear in the paper.

Accepted electronically prepared files can be submitted via the web at `www.ams.org/submit-book-journal/`, sent via FTP, or sent on CD-Rom or diskette to the Electronic Prepress Department, American Mathematical Society, 201 Charles Street, Providence, RI 02904-2294 USA. TeX source files, DVI files, and PostScript files can be transferred over the Internet by FTP to the Internet node `ftp.ams.org` (130.44.1.100). When sending a manuscript electronically via CD-Rom or diskette, please be sure to include a message identifying the paper as a Memoir.

Electronically prepared manuscripts can also be sent via email to `pub-submit@ams.org` (Internet). In order to send files via email, they must be encoded properly. (DVI files are binary and PostScript files tend to be very large.)

Electronic graphics. Comprehensive instructions on preparing graphics are available at `www.ams.org/authors/journals.html`. A few of the major requirements are given here.

Submit files for graphics as EPS (Encapsulated PostScript) files. This includes graphics originated via a graphics application as well as scanned photographs or other computer-generated images. If this is not possible, TIFF files are acceptable as long as they can be opened in Adobe Photoshop or Illustrator. No matter what method was used to produce the graphic, it is necessary to provide a paper copy to the AMS.

Authors using graphics packages for the creation of electronic art should also avoid the use of any lines thinner than 0.5 points in width. Many graphics packages allow the user to specify a "hairline" for a very thin line. Hairlines often look acceptable when proofed on a typical laser printer. However, when produced on a high-resolution laser imagesetter, hairlines become nearly invisible and will be lost entirely in the final printing process.

Screens should be set to values between 15% and 85%. Screens which fall outside of this range are too light or too dark to print correctly. Variations of screens within a graphic should be no less than 10%.

Inquiries. Any inquiries concerning a paper that has been accepted for publication should be sent to `memo-query@ams.org` or directly to the Electronic Prepress Department, American Mathematical Society, 201 Charles St., Providence, RI 02904-2294 USA.

Editors

This journal is designed particularly for long research papers, normally at least 80 pages in length, and groups of cognate papers in pure and applied mathematics. Papers intended for publication in the *Memoirs* should be addressed to one of the following editors. The AMS uses Centralized Manuscript Processing for initial submissions to AMS journals. Authors should follow instructions listed on the Initial Submission page found at www.ams.org/memo/memosubmit.html.

Algebra to ALEXANDER KLESHCHEV, Department of Mathematics, University of Oregon, Eugene, OR 97403-1222; email: ams@noether.uoregon.edu

Algebraic geometry to DAN ABRAMOVICH, Department of Mathematics, Brown University, Box 1917, Providence, RI 02912; email: amsedit@math.brown.edu

Algebraic geometry and its applications to MINA TEICHER, Emmy Noether Research Institute for Mathematics, Bar-Ilan University, Ramat-Gan 52900, Israel; email: teicher@macs.biu.ac.il

Algebraic topology to ALEJANDRO ADEM, Department of Mathematics, University of British Columbia, Room 121, 1984 Mathematics Road, Vancouver, British Columbia, Canada V6T 1Z2; email: adem@math.ubc.ca

Combinatorics to JOHN R. STEMBRIDGE, Department of Mathematics, University of Michigan, Ann Arbor, Michigan 48109-1109; email: JRS@umich.edu

Commutative and homological algebra to LUCHEZAR L. AVRAMOV, Department of Mathematics, University of Nebraska, Lincoln, NE 68588-0130; email: avramov@math.unl.edu

Complex analysis and harmonic analysis to ALEXANDER NAGEL, Department of Mathematics, University of Wisconsin, 480 Lincoln Drive, Madison, WI 53706-1313; email: nagel@math.wisc.edu

Differential geometry and global analysis to CHRIS WOODWARD, Department of Mathematics, Rutgers University, 110 Frelinghuysen Road, Piscataway, NJ 08854; email: ctw@math.rutgers.edu

Dynamical systems and ergodic theory and complex analysis to YUNPING JIANG, Department of Mathematics, CUNY Queens College and Graduate Center, 65-30 Kissena Blvd., Flushing, NY 11367; email: Yunping.Jiang@qc.cuny.edu

Functional analysis and operator algebras to DIMITRI SHLYAKHTENKO, Department of Mathematics, University of California, Los Angeles, CA 90095; email: shlyakht@math.ucla.edu

Geometric analysis to WILLIAM P. MINICOZZI II, Department of Mathematics, Johns Hopkins University, 3400 N. Charles St., Baltimore, MD 21218; email: trans@math.jhu.edu

Geometric topology to MARK FEIGHN, Math Department, Rutgers University, Newark, NJ 07102; email: feighn@andromeda.rutgers.edu

Harmonic analysis, representation theory, and Lie theory to ROBERT J. STANTON, Department of Mathematics, The Ohio State University, 231 West 18th Avenue, Columbus, OH 43210-1174; email: stanton@math.ohio-state.edu

Logic to STEFFEN LEMPP, Department of Mathematics, University of Wisconsin, 480 Lincoln Drive, Madison, Wisconsin 53706-1388; email: lempp@math.wisc.edu

Number theory to JONATHAN ROGAWSKI, Department of Mathematics, University of California, Los Angeles, CA 90095; email: jonr@math.ucla.edu

Number theory to SHANKAR SEN, Department of Mathematics, 505 Malott Hall, Cornell University, Ithaca, NY 14853; email: ss70@cornell.edu

Partial differential equations to GUSTAVO PONCE, Department of Mathematics, South Hall, Room 6607, University of California, Santa Barbara, CA 93106; email: ponce@math.ucsb.edu

Partial differential equations and dynamical systems to PETER POLACIK, School of Mathematics, University of Minnesota, Minneapolis, MN 55455; email: polacik@math.umn.edu

Probability and statistics to RICHARD BASS, Department of Mathematics, University of Connecticut, Storrs, CT 06269-3009; email: bass@math.uconn.edu

Real analysis and partial differential equations to DANIEL TATARU, Department of Mathematics, University of California, Berkeley, Berkeley, CA 94720; email: tataru@math.berkeley.edu

All other communications to the editors should be addressed to the Managing Editor, ROBERT GURALNICK, Department of Mathematics, University of Southern California, Los Angeles, CA 90089-1113; email: guralnic@math.usc.edu.

Titles in This Series

941 **Gelu Popescu,** Unitary invariants in multivariable operator theory, 2009

940 **Gérard Iooss and Pavel I. Plotnikov,** Small divisor problem in the theory of three-dimensional water gravity waves, 2009

939 **I. D. Suprunenko,** The minimal polynomials of unipotent elements in irreducible representations of the classical groups in odd characteristic, 2009

938 **Antonino Morassi and Edi Rosset,** Uniqueness and stability in determining a rigid inclusion in an elastic body, 2009

937 **Skip Garibaldi,** Cohomological invariants: Exceptional groups and spin groups, 2009

936 **André Martinez and Vania Sordoni,** Twisted pseudodifferential calculus and application to the quantum evolution of molecules, 2009

935 **Mihai Ciucu,** The scaling limit of the correlation of holes on the triangular lattice with periodic boundary conditions, 2009

934 **Arjen Doelman, Björn Sandstede, Arnd Scheel, and Guido Schneider,** The dynamics of modulated wave trains, 2009

933 **Luchezar Stoyanov,** Scattering resonances for several small convex bodies and the Lax-Phillips conjuecture, 2009

932 **Jun Kigami,** Volume doubling measures and heat kernel estimates of self-similar sets, 2009

931 **Robert C. Dalang and Marta Sanz-Solé,** Hölder-Sobolv regularity of the solution to the stochastic wave equation in dimension three, 2009

930 **Volkmar Liebscher,** Random sets and invariants for (type II) continuous tensor product systems of Hilbert spaces, 2009

929 **Richard F. Bass, Xia Chen, and Jay Rosen,** Moderate deviations for the range of planar random walks, 2009

928 **Ulrich Bunke,** Index theory, eta forms, and Deligne cohomology, 2009

927 **N. Chernov and D. Dolgopyat,** Brownian Brownian motion-I, 2009

926 **Riccardo Benedetti and Francesco Bonsante,** Canonical wick rotations in 3-dimensional gravity, 2009

925 **Sergey Zelik and Alexander Mielke,** Multi-pulse evolution and space-time chaos in dissipative systems, 2009

924 **Pierre-Emmanuel Caprace,** "Abstract" homomorphisms of split Kac-Moody groups, 2009

923 **Michael Jöllenbeck and Volkmar Welker,** Minimal resolutions via algebraic discrete Morse theory, 2009

922 **Ph. Barbe and W. P. McCormick,** Asymptotic expansions for infinite weighted convolutions of heavy tail distributions and applications, 2009

921 **Thomas Lehmkuhl,** Compactification of the Drinfeld modular surfaces, 2009

920 **Georgia Benkart, Thomas Gregory, and Alexander Premet,** The recognition theorem for graded Lie algebras in prime characteristic, 2009

919 **Roelof W. Bruggeman and Roberto J. Miatello,** Sum formula for SL_2 over a totally real number field, 2009

918 **Jonathan Brundan and Alexander Kleshchev,** Representations of shifted Yangians and finite W-algebras, 2008

917 **Salah-Eldin A. Mohammed, Tusheng Zhang, and Huaizhong Zhao,** The stable manifold theorem for semilinear stochastic evolution equations and stochastic partial differential equations, 2008

916 **Yoshikata Kida,** The mapping class group from the viewpoint of measure equivalence theory, 2008